VISUALIZING YOUR BUSINESS

Let Graphics Tell the Story

VISUALIZING YOUR BUSINESS

Let Graphics Tell the Story

Keith R. Herrmann

WILEY

John Wiley & Sons, Inc.

New York • Chichester • Weinheim • Brisbane • Singapore • Toronto

Library of Congress Cataloging-in-Publication Data
Herrmann, Keith R.
    Visualizing your business : let graphics tell the story / Keith R. Herrmann.
      p. cm.
    Includes index.
    ISBN 0-471-37199-8 (cloth : alk. paper)
    1. Business presentations—Graphic methods. I. Title
   HF5718.22 .H47 2001
   658.4'5—dc21
                                                       00-068487

Printed in the United States of America.

10 9 8 7 6 5 4 3 2 1

For Brenda:
Were it not for her, the book would
never have been written and
my life would be unimaginably dimmer.

"Keith Herrmann's book should be required reading for managers at all levels."

—Frederick Smith
Chairman and CEO, FedEx Corporation

"Keith Herrmann has provided business executives with an indispensable guide for exhibiting data with clarity and effectiveness to ensure successful business presentations."

—Burton Malkiel
Professor, Princeton University
Author of *A Random Walk Down Wall Street*

"My passion has always been to make large quantities of investment data accessible to everyone. Keith Herrmann's new book provides 101 ways to better communicate business information. It's fun and easy to read."

—Roger G. Ibbotson
Professor, Yale School of Management
Chairman, Ibbotson Associates

"Mr. Herrmann has done it again: taken a seemingly simple task, preparing (and, of course, delivering) charts and graphs, and turned it into a learned if entertaining treatise—as usual, with a deft touch and an intimate knowledge of how thirsting for knowledge CEOs and other corporate executives are. He has turned a simple subject into a minor masterpiece. All you never knew you needed to know concerning charts and graphs and their preparation."

—Sid Cato
President of Cato Communications
Publisher of *Newsletter on Annual Reports*

"Mr. Herrmann addresses an increasingly important communication area with clarity and breadth."

—Gary DiCamillo
Chairman and CEO, Polaroid Corporation

"Herrmann's book is like a Swiss army knife; with it, you will be prepared for just about any contingency."

—Mark Baric
President and CEO, Virtus Corporation

Acknowledgments

I would like to acknowledge and thank the many professionals who provided both guidance and support along the way, and in particular all those who reviewed the manuscript and gave valuable comments and suggestions: David Axson, Mort Barlaz, Lester Bumas, Mark Herrmann, Alan Kritz, Denny LeSage, Ellen Levine, Bill Raddi, and Terry Ribb. I consider myself fortunate to have mentors and friends of their caliber.

K.R.H.

Foreword

I first got to know Keith Herrmann some years ago when he began submitting articles to my publication, *Across the Board*, the management magazine published by The Conference Board. We get many over-the-transom submissions, and usually—and for sound reason—they get short shrift. But despite initial rejection, Keith kept at it, and his pieces got longer and longer shrift, and eventually we began publishing his funny, pun-filled articles.

Some of his puns were good, some bad, and some were so bad they were good. Puns are a long way from charts and graphs, a long way from corporate finance, which is where Keith's professional expertise lies. But perhaps not: Puns demand a playful mind, one that is engaged by topsy-turvy (and what are the best charts and graphs but imaginative reframings of what can be some pretty dull stuff?); and corporate finance, of course, demands an engagement with numbers (or Numbers, as they are regarded fearfully by those of us who are mathematically challenged).

Which brings me to a major point: that the charts and graphs that illustrate this volume are understandable—yea, even among the most benighted among us. Which is not to say that Keith's 101 charts should be characterized as Charts 101. Yes, there is a lot of basic material here, but there is also stuff for those looking to take the advanced course.

Happily, too, while Keith is utterly taken with his subject, he is not fanatic about it. Most business book writers imply that their books will make you a better person, get you a raise in salary, get you a promotion, or even put you in the chief executive's chair. Keith does not make those kinds of promises. Using charts and graphs well, he suggests, will enable you to communicate more effectively—to organize and thus simplify complexity—but the rest is up to you.

While this is a serious book on a serious subject, Keith's puckish humor inevitably pokes through. An example: See his explanation, in Chapter 11, of why a brigh-line chart is called a bright-line chart (when it is colored gray).

Humor aside, the book goes to present solutions to what are two very real problems: (1) we are overwhelmed with data and information, and (2) there is a desperate need to make ourselves understood. Keith's 101 charts and graphs are not perfect solutions to these problems—unfortunately, there are none. But they are sensible and significant steps toward that ideal.

A. J. Vogl
Editor, *Across the Board* magazine

About the Author

Keith R. Herrmann has held key strategic planning and financial analysis positions with Black & Decker, Exide Electronics Group, Invensys Plc, and Midway Airlines. He is the author of numerous articles, and his work has appeared in the *Journal of Accountancy*, *PC World*, *Strategic Finance*, and the *Journal of Construction Accounting and Taxation*. He has been featured at national conferences, speaking on how to present ideas clearly. Mr. Herrmann was educated at the University of Pennsylvania; King's College London; the Sorbonne; and the Yale School of Management. He is interested in receiving any examples you might have of excellent business charts and graphs. He can be reached at 919-848-0323; e-mail: krherrmann@aol.com.

Preface

Welcome to *Visualizing Your Business: Let Graphics Tell the Story*. This book is different from most others on charts and graphs that you might have seen, because it focuses on a topic that is important to all of us: business.

Why I Wrote This Book

When I first began working as a financial analyst, I wanted to learn as much as possible to excel in my career. Having majored in both math and English, I thought that the demands placed on a financial manager required a deeper understanding of the numbers. I looked for a book to guide me when presenting complex business issues. To my surprise, no such book existed. This book fills that gap.

I set out to write a book that practitioners engaged in any level of business could use, from basic financial concepts such as sales and cash flow, to operational tools such as quality control, to investment decisions such as business case analysis. This book is not just for the entry-level analyst who might be creating the graphics but also for the recipients of that information.

Overview of This Book

Each chapter in *Visualizing Your Business: Let Graphics Tell the Story* provides information about a different business task or procedure and discusses the best charts and graphs to apply to that situation. You will find examples that you may already use in your business activities on a daily basis. But you will also find tools that you have never used or that you might not have considered using in that particular context. After all, no one can be completely familiar with every type of chart and graph. So we all have gaps in our knowledge. The purpose here is to help fill in some of the gaps that may have developed since your last course in business.

Visualizing Your Business: Let Graphics Tell the Story makes liberal use of case studies: that is, situations that are typical of decisions or problems that you might face on any given workday. The case studies demonstrate at least one possible graphical solution as a tool. The intent is for you to mentally put yourself in the situation described, work through it, and then apply the solution to an actual situation that you face.

How to Use This Book

You can dip into this book to find particular topics and to make use of the information without necessarily reviewing everything that came before. With access to the 101 different types of charts in this book, you can pick the correct chart for a given situation. If you are trying simply to navigate data overload, then this book will help you to interpret reams of data. *Visualizing Your Business: Let Graphics Tell the Story* is intended to become your trusted guide as you navigate the ghastly complexity of business. It will help make the difficult, often impossible job of analyzing and presenting complex ideas and concepts to clients and associates fast, simple, thorough, and inexpensive.

Many of the charts and graphs could have been included in more than one section, but they are classified according to their dominant characteristic. For example, the Marimekko graph is categorized in the "Variances and Comparisons" chapter, but it could have been included in the "Tracking" or "Relationships" chapters as well. As you become familiar with these charts and graphs, you will begin to use them in different settings.

Ultimately, the value of charting to managers lies as much in the process of creating the chart as in the consumption of its product. Charting stimulates the development of a deeper understanding of a business and its environment, and it forces the formulation and evaluation of alternatives that would not otherwise be considered. It unleashes large amounts of creativity that are so often suppressed by routine and the need to respond to crises.

How This Book Is Organized

You can look in the table of contents or the index of *Visualizing Your Business: Let Graphics Tell the Story* for all situations, whether familiar or unfamiliar, and read about how to analyze them by means of the tools. To make it easier to find related situations, the book is divided into 16 chapters plus three appendices.

Chapter 1, "Introduction," discusses where to start, when to use graphics, and how to get numbers to tell the story.

Chapter 2, "Tracking," contains eight charts for tracking fundamental financial concepts such as sales, cash flow, and income statements. The use of these charts will provide you with an overview of your business.

Chapter 3, "Variances and Comparisons," covers eight techniques to display systems of interrelated values so that deviations, groupings, and the relative sizes of more than one element can be seen at the same time.

Chapter 4, "Trends and Change," presents 13 ways to depict trends, change, and quality control procedures.

Chapter 5, "Relationships," contains 12 charts and graphs, including supply and demand, Venn diagrams, and other ways to show relationships.

Chapter 6, "Presentation," covers 10 techniques that have a wide variety of applications.

Chapter 7, "Value Ranges," presents confidence intervals, high-low graphs, and candlestick charts.

Chapter 8, "Schematics and Maps," covers schematics, process diagrams, and maps.

Chapter 9, "Organizing Data and Information," presents structure charts and other ways to organize data.

Chapter 10, "Planning, Scheduling, and Project Management," covers time charts, including time lines, PERT charts, and Gantt charts.

Chapter 11, "Probability, Prediction, and What-If," covers decision trees, payoff tables, and bright-line charts.

Chapter 12, "Strategy," discusses life cycle analysis, growth/share matrix, and value chain. How to structure, test, and quantify strategies are covered here, as well as how to visualize your competitive advantages and disadvantages.

Chapter 13, "Finance," contains case studies of how to present more detailed financial concepts.

Chapter 14, "Marketing," contains case studies of how to present sales and marketing analysis and market positioning.

Chapter 15, "Performance," contains case studies of how to present various other business performance measurements.

Chapter 16, "A Thousand Points of Data," includes a brief summary.

Appendices discuss the anatomy of a chart, defining chart types, and financial charting in Excel 97.

A CD accompanies this book. For your convenience, the CD contains Excel and PowerPoint files that carry 25 of the tools used in *Visualizing Your Business: Let Graphics Tell the Story*. Throughout the book, the following symbol (●) is placed next to each of the tools that is included on the CD.

About the CD-ROM

Introduction

The files on the enclosed CD-ROM are saved in Microsoft Excel and PowerPoint for Windows. In order to use the files, you will need to have spreadsheet software capable of reading Microsoft Excel for Windows and PowerPoint files.

System Requirements

- IBM PC or compatible computer
- CD-ROM drive
- Windows 95 or later
- Microsoft Excel for Windows or other spreadsheet software capable of reading Microsoft Excel for Windows files
- Microsoft PowerPoint for Windows

Note: Many popular spreadsheet programs are capable of reading Microsoft Excel 97 for Windows files. However, users should be aware that a slight amount of formatting might be lost when using a program other than Microsoft Excel 97.

If you do not have Microsoft PowerPoint 97, you can download the Microsoft Viewer for free from Microsoft's website:

(http://officeupdate.microsoft.com/2000/downloadDetails/Ppview 97.htm)

Using the Files

Loading Files

To use the spreadsheet, launch your spreadsheet program. Select File, Open from the pull-down menu. Select the appropriate drive and directory. Double-click on the file you want to open. Edit the file according to your needs.

To use the PowerPoint files, launch the Microsoft PowerPoint program. Select Open an existing presentation from the dialog box. Select the appropriate drive and directory. Double click on the presentation you want to open.

Printing Files

If you want to print the files, select File, Print from the pull-down menu.

Saving Files

When you have finished editing a file, you should save it under a new file name by selecting File, Save As from the pull-down menu.

User Assistance

If you need assistance with installation or if you have a damaged CD-ROM, please contact Wiley Technical Support at:

Phone: (212) 850-6753

Fax: (212) 850-6800 (Attention: Wiley Technical Support)

Email: techhelp@wiley.com

To place additional orders or to request information about other Wiley products, please call (800) 225-5945.

Contents

1. Introduction

A chief executive officer (CEO) is floating in a hot-air balloon. He's lost. He maneuvers the balloon down closer to the ground to find out where he is. He sees a man in a field and floats over to him, about 40 feet off the ground.

"Excuse me," he calls, "Can you tell me where I am?"

The guy looks up: "Sure. You're in a balloon. Forty feet off the ground. Over a field."

The CEO looks down and shouts: "You must be a consultant. You gave me data that is technically correct but isn't useful."

The guy on the ground looks up and exclaims: "And you, sir, must be a CEO."

"Why do you say that?" asks the perplexed CEO.

The guy on the ground says: "Because you have no idea where you are. No idea where you are going. And you assume it's all my fault!"

Like the CEO in the hot-air balloon, many of today's business leaders are lost. They have no idea where they are and no idea where they are going. They fall under the delusion that mere data is enough. If we have the facts, then what more do we need?

Business leaders are like artists during the Renaissance. The Renaissance artist might take a commission for a portrait one day, carve a tomb the next, and then design a palace or work out a problem in military engineering. Business leaders, who must run all aspects of a company, are asked to do the same.

A superior individual in the Renaissance, however, could assemble in his or her own head all the significant data of the time. Because of the extreme quantity of data and the diversity of the problems to which we are exposed, a businessperson in the twenty-first century cannot possibly do this. A 10-year-old child today has more data available than Leonardo da Vinci had at his prime. A small business or industry is already too complex to be handled by a single individual. The problem is how to cope effectively with increasing quantities of data.

This book seeks to enlarge your capacity to process the vast quantities of data that you possess. The book is a treasury of the best 101 examples of business graphics. If you are lost like the CEO in the hot-air balloon, then it will help you become a more skillful navigator. If you are trying simply to navigate data overload, then it will help you to interpret reams of data. You need informative and eye-appealing charts and graphs to make data meaningful. You can see in graphic form what was not so clear before and pull out that nugget you might have missed that will make your business decision the right decision.

This book, however, is not for studying "everything." Delineating a clear direction is a prerequisite for success. Knowing when to use the available tools is also a key to success, and there is no substitute for the hard work of selecting and using these tools to craft solutions tailored to your company's unique context and needs. We will tear ourselves to pieces, trying to go in 100 different directions at the same time, unless we find out what we want most and make that the goal of our supreme effort. Demonstrating leadership requires the willingness to assess situations, think through the available options, decide on the tools to be used and how to adapt them as needed, and then accept accountability for the decisions made and the results achieved.

I have tried to address those questions about charts and graphs that managers seem to be most interested in and to do so in a way that provides answers that are both understandable and useful. One of the myriad challenges I faced when writing this book was to address the needs of my readers from the perspective of the skill sets that they already possess. This proved to be a delicate balancing act: not only to provide the neophyte with a sweeping portrait of charts and graphs but also to give the more experienced reader additional insight from these charts. I wish neither to confuse the novice with technicalities nor to insult the seasoned professional with platitudes. This book is not intended as a scholarly text, still less an exhaustive one. For those of you who may already be familiar with certain charts and graphs included here, I hope that you understand the need for their inclusion. Whenever I mention a business concept (such as return on investment or statistical process control), I have also tried to include a reference. Consult the reference material if you are unfamiliar with these terms or are interested in finding out more about them.

This book is an introduction to its subject for those who feel the same curiosity about it that I do. Charts and graphs have to be tailored to the unique characteristics of the organization and situation for which they are made. My intent is for each chart or graph to be modified and used as needed, at whatever level of complexity, in the manner that is most appropriate to the task at hand.

Finding the Graphics You Need

To make it easier to find related graphics, the book is organized into 16 chapters plus three appendices. Many of the charts and graphs could have been included in more than one section, but they are classified according to their dominant characteristic.

- "Introduction." This chapter discusses where to start, when to use graphics, and how to get numbers to tell the story.

- "Tracking." This chapter contains eight charts for tracking fundamental financial concepts such as sales, cash flow, and income statements. The use of these charts will provide you with a business overview.

- "Variances and Comparisons." This chapter covers eight techniques to display systems of interrelated values so that deviations, groupings, and the relative sizes of more than one element can be seen at the same time.

- "Trends and Change." You will find 13 ways to depict trends, change, and quality control procedures.

- "Relationships." This chapter contains 12 charts and graphs, including supply and demand, Venn diagrams, and other ways to show relationships.

- "Presentation." Ten techniques that have a wide variety of applications are covered here.

- "Value Ranges." This chapter covers confidence intervals, high-low graphs, and candlestick charts.

- "Schematics and Maps." This chapter covers schematics, process diagrams, and maps.

- "Organizing Data and Information." Structure charts and other ways to organize data are presented here.

- "Planning, Scheduling, and Project Management." This chapter covers time charts, including time lines, PERT charts, and Gantt charts.

- "Probability, Prediction, and What-If." This chapter covers decision trees, payoff tables, and bright-line charts.

- "Strategy." You will find life cycle analysis, growth/share matrix, and value chain in this chapter. How to structure, test, and quantify strategies are covered here, as well as how to visualize your competitive advantages and disadvantages.

- "Finance." This chapter contains case studies of how to present more detailed financial concepts.

- "Marketing." This chapter contains case studies of how to present sales and marketing analysis and marketing positioning.

- "Performance." This chapter contains case studies of how to present various other business performance measurements.

- "A Thousand Points of Data." This chapter includes a brief summary.

- "Appendices." The appendices discuss the anatomy of a chart, defining chart types, and financial charting in Excel 97.

The tools in this book will help you to set guidelines and measure progress. They will help you to convey your company's story, and they will reveal your company's location and orientation. They will help you to set your course and to select your destination. And they will help you to arrive at your destination more quickly. While some of these tools may already be familiar to you, others will be new. What is not new is the need for the courage to manage: to assess situations, set an overall course or focus, think through options, develop plans, take action, modify plans, learn, and go forward.

Moving at the Speed of Business

Today, more than ever, companies have to work harder to keep up with the pace of change and increased global competition. To succeed in multi-channel, high-speed environments, you need to leverage the data you have at your disposal. You need to harness the knowledge inherent in your organization—at both the individual and the corporate levels. These exercises can be vital to your organization's survival, because helping managers understand and remember complex data can give your organization a competitive edge. The tools in this book will help managers deal with the ever-increasing volumes and types of internal and external data. Using effective charts and graphs enables the knowledge worker to become more productive and effective, thus increasing both personal and corporate knowledge.

From the shop floor to the boardroom, everybody talks about the speed at which things are changing. We hear a lot about the emergence of new technologies that render entire markets obsolete overnight; about the increased speed of communication, which allows shock to travel instantaneously across the planet; and about the shifting of boundaries between industries and the emergence of new competitors.

While there is nothing new about the inevitability of change, there is perhaps a new way of thinking about the rate, or speed, of change. Consider the advances in one area of scientific accomplishment: the basic acceleration of travel as measured against time. In 1520, Magellan's crew circumnavigated the globe in a wood-hulled ship. About 350 years later, people circumnavigated the globe in a steel steamship; 60 years after that, in an airplane; and after another 35 years, in a space capsule. To circumnavigate the globe, the wood-hulled ship took two years, the steel steamship two months, the airplane two weeks, and the space capsule took only a little over one hour. This example contains not one but two dimensions of acceleration. The first acceleration is the contraction of the lags between the successive circumnavigations from 350 years to 60 years to 35 years. The time that elapsed between the four states of the art of circumnavigation grew shorter each time. The second acceleration is the contraction of time taken to actually circumnavigate the globe from two years...to two months...to two weeks...to one hour.

Likewise in business, there are (at the very least) two types of acceleration: Time to market is shrinking and product life cycles are collapsing. Consider the change in the amount of time it took to launch a new car or truck at the beginning of the twenty-first century. In 1990, it took U.S. manufacturers six years to go from concept to production of an automobile; now it takes just two years to bring automotive products to market. Remember when it took weeks to get eyeglasses? Now it takes under an hour.

In a similar fashion, product life cycles are shrinking. It is difficult to predict where the marketplace is going in the short term, and no one knows what the best-selling products will be one year from now. Because product sales are hard to predict, many companies have abandoned traditional annual budgets in favor of budgets with shorter time horizons. For example, the budgets at Nortel extend only for the next six months. In the telecommunications industry, it is more than likely that the cell phone that was "cutting edge" yesterday has become obsolete virtually overnight.

The new enterprise needs to be a real-time business, making adjustments now to changes in market conditions and consumer needs. Wait, and by the time you get to market, it will be too late. Conditions will have changed, and customers will be demanding new and different products. This is an environment that favors agile businesses that can respond quickly to the marketplace.

Your company needs to be fast, and your company needs to be faster than that. Why? Because your customers demand it. Customers are no longer satisfied with the old paradigm of "cheaper, better, and faster"; they literally want your products and services "free, perfect, and now." You can improve your agility by improving the clarity of your analysis and presentations with meaningful charts and graphs. Using the types of charts and graphs in this book will help you to gain insights faster.

Helping your company managers understand your data will enable them to make better decisions and to perform their jobs more efficiently and effectively.

What Gets Measured Gets Done

From the point of view of business leadership, what matters is understanding how people react in times of uncertainty. During these times, managers have less ability to control their external environment. However, when managers have a framework within which to measure and implement change, then these focused measurements put change in perspective. As conditions change, focused measurements can enable a company to move both quickly and assuredly in order to seize new opportunities. How can your company begin to use focused measurements? To answer this question, we first need to understand what exactly focused measurements are.

One example of focused measurements is a baseball scoreboard. For example, at the end of the last game of the 2000 World Series, the scoreboard read:

	1	2	3	4	5	6	7	8	9	Runs	Hits	Errors
New York Yankees	0	1	0	0	0	1	0	0	2	4	7	1
New York Mets	0	2	0	0	0	0	0	0	0	2	8	1

By looking at the scoreboard, we know that the Yankees beat the Mets. If the measurements for the game were unfocused, then the scoreboard might look something like this:

Number of hot dogs served at concession stands	Number of wild pitches that hit the dirt	Number of times the lead changed hands
55,292	0	3

These measurements would make ridiculous headlines in the sports section of the newspaper the day after this game. Even though these statistics may be unique to this game, the fans do not need to know them. What the fans want and expect is the number of runs, hits, and errors, and then the detail on the runs, inning by inning. That is why the baseball scoreboard is such an effective tool.

Perhaps the runs-hits-errors approach is not to be held up as an ideal—it is not data rich, while we have the potential to generate great graphics of baseball nowadays. Instant replay boards in stadiums display a lot of graphics now; the current baseball scoreboard is more vibrant and data rich than the old-fashioned one. TV coverage has rich graphics of baseball.

The runs, hits, and errors are the vital statistics that create uniformity among the outcomes of the games and enable the tabulation of results. Each team needs these measurements to record the results of the game and to participate in baseball league play. So, too, do public companies need certain numbers, reports, and information for the shareholders, stock exchanges, regulatory agencies, and others. Moreover, what both baseball teams and businesses look at and measure individually will help determine the future numbers and the statistics for improved performance over the next season or quarter. And both baseball teams and businesses have to make choices even though they cannot know, with complete certainty, where they will end up as a consequence of the actions taken today.

The reporting of business performance is similar to the use of a scoreboard to report on baseball performance; both are at the tail end of a process that begins with defining information needs. Who won the game? What were the quarterly earnings per share? The reporting can begin only after the appropriate measurements have been determined. To extend this analogy just a little bit further, defining information needs in business is often a race without a finish line, since businesses are constantly changing. Change makes the information needs of businesses dynamic because what is important to them changes with time. For example, a start-up company may begin by measuring customer acquisition; as the company matures, measuring customer retention becomes more appropriate. The questions are moving targets. Providing the right information in a timely fashion requires anticipation.

Why Use Graphics?

This book uses many kinds of charts and graphs to focus measurements and to enable managers to unlock the potential hidden in their business data. Graphics for business purposes are used daily by millions of people for such things as:

- Improving efficiency and effectiveness
- Improving quality
- Solving problems

- Planning
- Teaching
- Training
- Monitoring processes
- Looking for trends and relationships
- Reviewing the status of projects
- Developing ideas
- Writing reports
- Studying sales results
- Reducing costs

With ever-increasing amounts of data, businesses often use charts and graphs to increase productivity and improve quality. Fortunately, as a result of developments in computer equipment and software, most of the popular charts and graphs used on a daily basis can be generated rapidly, easily, and with little or no special training. Yet the choice of which charts and graphs to use is still an arbitrary process.

An important note: The recommendations offered in this book are guidelines, not hard and fast rules. In some cases, aesthetic concerns, context, or your intended emphasis may override my advice. Moreover, the recommendations occasionally may interact in unexpected ways. Nevertheless, some guidance at a fairly general level is possible. It is sometimes helpful to generate a rough sketch of several alternative displays that follow the recommendations and then to evaluate each one. This is particularly easy to do on a personal computer with a good graphics program.

Many of the charts and graphs presented here were created using the Microsoft Office Suite of software (PowerPoint and Excel). The CD included with this book has templates that will save you much of the initial time required to set up the graphics; then these formats can be used multiple times. Subsequent use is much less time-consuming.

Many books are available on computer graphics. These other books tend to focus on topics such as websites; three-dimensional graphics; geometry and mathematics; engineering; or how to use a specific software package, such as CorelDraw or Freelance Graphics. None of these other books combine management and graphical presentations in the way that *Visualizing Your Business: Let Graphics Tell the Story* does.

Where to Start

Clarity is the first aim, economy the second, grace the third.

—Sheridan Baker

All too often great analysis is lost in a fog of spreadsheets. The ability to present sound analysis with clarity and insight distinguishes great analysts from good number crunchers. I believe that the ability to produce good charts and graphs is the result of thinking about an issue as a complex network of problems. Charting is therefore both a *science* and an *art*. As a *science,* it is a tool to analyze problems. As an *art*, it helps to coordinate the activities of other employees. It is one way to describe an issue from a particular point of view at a particular point in time. The best charting of which we are capable requires at least as much science as it does art. I am as interested in improving the art as I am in improving the science.

Most managers agree: It is easy to get data, but it is increasingly difficult to convert this data into meaningful information. Managers today complain of "drowning in data while thirsting for information." Data is everywhere, but it never seems to be what the managers want. Most database reports get thrown away without ever being looked at. How can that be? The answer is simple: *Data* and *information* are not the same thing! There is a need to reduce the vast array of data to a smaller set of information that is useful for decision making and understanding. Information *tells* you something. It informs you. That is the key characteristic that makes information valuable to managers.

You might say that a manager's use of a chart to present a business issue is not unlike what a movie director, using a technique such as montage, might do with something called "boy meets girl." Or what the super high-speed camera does with a flame at the instant that it is snuffed out. The clarity being sought is achieved only by accepting increasing complexity and establishing a common framework within which issues can be related and evaluated. If things never changed, then getting all the right information to all the right people in any organization would be a simple matter of time. Eventually all the important data would be available, and all the important views of the business would be known. But the reality is that business is dynamic. No matter what business you are in, nothing stays the same for long. Change leads to competition, competition fuels ingenuity, and ingenuity drives change. New products and services get introduced. Old ones become obsolete. New opportunities pop up—along with new competitors. Managers reorganize to align with new market realities. New processes replace old ones. This cycle of change presents opportunities to those willing to rise to the challenge.

A chart is a special kind of picture that has no visible resemblance to the physical world. It is an abstraction, in much the same way that letters and numbers are abstractions. And just as there are 26 letters in the alphabet and 10 Arabic numerals, there are also a finite number of chart types that can be counted and indexed.

I have found that even though businesses create an enormous and perishable mass of data, limited numbers of enduring chart types are used to present that data. Businesses tend to convey similar issues over and over again using the same types of charts and graphs. While you may have used only a fraction of these charts, the larger your portfolio of charts, the more likely it is that you will find a chart that is ideally suited to convey your message.

The huge volume of information generated as a by-product of managing an enterprise demands a pragmatic new approach to filtering out the nonessential data and correlating one piece of data with another. Using graphics is a technique for dealing with the dual demands of handling complexity while simultaneously providing simplicity. Graphics are abstract pictures that can communicate the information contained within numbers to allow a better understanding of the numbers and therefore enable better decisions. For this book to be beneficial, the proper (or most useful) graphic must be selected. Even after selecting the type of graph, a large number of decisions related to the actual representation must be made. To optimize the effect of a graphic, you must look at the details of a representation and ensure that the data is revealed rather than obscured.

A primary purpose of an accounting system is to accumulate and to communicate economic data for use by those making decisions. Due to the proliferation of computers, over the past several years there has been an explosion of information available to these decision makers. There has also been a shift in organizational culture toward increased employee involvement in decisions. This means that there is an increase in the number of nonfinancial people involved in the organization's decision-making process. Many organizations are moving rapidly to situations in which more complex information is being interpreted and acted upon by decision makers without extensive financial experience.

Graphics have become an essential element in the communications arsenal of the modern business. Internally or externally, powerful graphics can send an unambiguous message. The best graphics are not puzzles to be pondered. They gracefully display a well-developed idea and well-researched numbers or facts. Virtually anyone can read them immediately. They help make people care about the point you are trying to make.

When to Use Graphics

A picture is worth ten thousand words.

—Chinese proverb

Nearly any kind of business will benefit from some form of chart or graph. Shareholders will have a better grasp of earnings or losses when a chart shows them where the money went. A vice president of operations will better understand his or her cost objective when a chart shows how the objective fits into the overall financial plan. People truly do stretch more when they can put their actions in the context of goals.

Charts and graphs are classified as *presentation* types when they are used primarily in *formal or semiformal presentations.* Graphics are often an essential component of a presentation. Charts and graphs are a "must" whenever it is essential that your audience retains your message and the presentation includes numbers or mathematical calculations. Another major classification is *operational* charts and graphs, used primarily for activities such as analyzing, planning, monitoring, decision making, and communicating in *the ongoing running of a business*. They are used to supplement or replace tabulated data and written reports. Both presentation and operational charts and graphs tend to expedite group decisions and shorten meeting time.

Did you know that most humans absorb more than 80 percent of what they learn through the sense of sight? That means that if you *show* something to people, they are far more likely to remember it, at least for a while, than if you *tell* something to them. *Show* and *tell* at the same time, and your audience will remember even more. People retain more information when they hear it and see it rather than just hear it. Two senses are better than one. But the chart must complement, not distract from, the oral presentation.

Even so, numbers do not easily yield their meaning when displayed in the typical spreadsheet format. Rows and columns communicate meaning poorly, because numbers are visually similar. In addition, relative position is a big aspect of visual communication, and yet it provides almost no information in a spreadsheet format. In a sea of numbers, grasping relationships between data points requires relatively detailed examination of each data point to convert the absolute value of the data points into their relative value. On the other hand, charts and graphs communicate meaning far more quickly and efficiently because the relative nature of the information is the most prominent feature. The exact value of any particular data point is visually subordinate to its relationship to other data points. In other words, charts more directly illustrate the context, which in turn provides meaning. No wonder that adding graphics to your presentations will increase the amount of information retained by your audience, and the use of graphics can reduce meeting time.

Geri McArdle, in *Delivering Effective Training Sessions*, writes that adding graphs, charts, maps, or photos to a presentation increases the amount of retained information by as much as 55 percent. For example, people who have attended a show-and-tell presentation will retain about 65 percent of the information after three days, compared to about 10 percent retention for audiences who have simply heard the information.

A study done by the Wharton School of Business demonstrated how the use of graphics reduced meeting times by as much as 28 percent. Another study found that audiences believe presenters who use graphics are more professional and credible than presenters who merely speak. And still other research indicates that meetings and presentations reinforced with visuals help participants reach decisions and consensus in less time. That's a strong case for translating numbers into graphics!

The following are the reasons for using presentation charts and graphs.

- They assist in the rapid orientation of an audience.
- They aid in the audience's understanding and retention of the material presented.
- The audience receives graphical material more favorably.
- The presenter is perceived as more prepared, professional, and interesting.

The following are the reasons for using operational charts and graphs.

- Data can be organized such that analysis is more rapid and straightforward.
- Viewers can more rapidly determine and absorb the essence of the information.
- Large amounts of facts can be more conveniently and effectively reviewed.
- Deviations, trends, and relationships stand out more clearly.
- Comparisons and projections often can be made more easily and accurately.
- Key concepts tend to be remembered longer.

To understand the power of charts and graphs, consider the story of how a chart helped change the placement of armor plating on Allied bombers during World War II. In this instance, a chart saved 1,000 Allied lives, maybe more. During the war, the mathematician Abraham Wald was trying to determine where to add extra armor to planes. He conducted painstaking research on the planes. Convinced that his hypothesis was correct, he plotted the pattern of bullet holes in returning aircraft. His conclusion was to determine carefully where returning planes had been shot and *put extra armor every place else!* Wald drew an outline of a plane and then put a mark on it where a returning aircraft had been shot. Soon the entire plane had been covered with marks *except* for a few key areas. Wald concluded that since planes had probably been hit more or less uniformly, those aircraft hit in the unmarked places had been unable to return. Thus those were the areas that required more armor. He presented his graphic to Air Force commanders, who added the armor. Wald helped prevent future Allied casualties by discovering how the planes had been shot down and making his case through the use of effective graphics.

Contrast Wald's wartime use of graphics to January 1986 when engineers at defense contractor Morton-Thiokol had a hunch that the space shuttle *Challenger* should not be launched because the cold weather might cause failure of the O-rings that helped seal joints on the rocket motors. To argue their point, they faxed 13 charts to NASA. However, according to informational graphics expert and Yale professor Edward Tufte, not one made a clear enough connection between cold temperatures and O-ring failures. The NASA decision makers were not convinced. The space shuttle was launched, the O-rings failed, *Challenger* exploded, and seven astronauts died.

Many graphics lack impact. While today's personal computer software packages, with their do-it-yourself graphics and programs to present images, have put graphics capabilities into many people's hands, these capabilities are often misused. By selecting the most meaningful tool from those presented in this book, charts and graphs can help make effective arguments in any profession—as long as the data is presented with both force and integrity. "There are displays that reveal the truth and those that do not," writes Tufte in *Visual Explanations.* The Allied bomber crews, the *Challenger* astronauts, and the CEO in the hot-air balloon all would agree that if the matter is an important one, then getting the displays of evidence right or wrong can possibly have momentous consequences.

Getting Numbers to Tell the Story

You've got to see it to believe it.

—Anonymous

Charts and graphs are often used to communicate focused messages. One of the most powerful ways to present numerical data is with charts and graphs—formats that can instantly translate an enormous collection of numbers instantly into concise, eye-appealing statements.

Here is a simple formula to remember when trying to get numbers to tell a story: Data + Relevance + Context = Information. Data is only one of the required ingredients in the recipe for information. To deliver information to decision makers, we must, at a minimum, provide *relevance* and *context,* in addition to data. In any given situation, only a limited number of the available data points are relevant. The first step in extracting information from data is sorting the relevant from the irrelevant. The next step is establishing context. Context is the multitude of important relationships that govern the interaction of one data point with another. Context is essential because decision makers are keenly interested in cause and effect and in where data points fit in relation to one another. Context provides meaning.

Adding charts and graphs to your presentations will increase the amount of information your audience retains. These graphics allow the audience to view only the data that is relevant to their immediate needs in the context within which they need to see it. Relevance and context are particularly troublesome for finance groups. The reason is that relevance and context depend on both the *individual* and the individual's particular *situation.* What is relevant to a chief financial officer may or may not be relevant to a vice president of marketing or a chief operations officer. Furthermore, because each has a different perspective, the desired information context could be different for each of them as well. To make matters worse, the decision makers themselves may not know ahead of time what data is relevant or what context they will need it in.

Few would argue that simply providing someone with information means that they automatically understand that information. Information is inanimate. Understanding, on the other hand, is the mental grasp or comprehension of that information. Obviously there is a difference, but what impact should that have on your choice of graphics?

If understanding is the objective, then it is important to realize that merely exposing decision makers to information does not necessarily provide the value that you are looking for. Information is not the primary end game, since it is one step further to insight. That is because comprehension is the final step in understanding. The information you make available has to make the journey into the mind of the decision maker and take up residence. Promoting understanding means that your graphics must support decision makers all the way from seeing data through comprehending it.

There are ways to maximize the probability that decision makers transform the information we provide into understanding. One very important step is optimizing the presentation, so its important meaning is communicated most effectively. The vast majority of business information is consumed visually. Whether in paper reports, spreadsheets, or memos, most of us depend on "seeing" to extract meaning. This bias results because our other senses (smell, taste, hearing, and touch) are seriously limited when it comes to communicating information that is largely numeric. The only efficient way to translate such complex information is visually.

Fortunately, with today's spreadsheet software, you do not have to undertake that translation manually. Most of the work is done for you with a few mouse clicks. For some charts, you may need to add a little formatting and a few formulas. The resulting graphic can then be printed or transmitted electronically.

The same data can be displayed in different ways and tell different stories. Consideration should be given to:

- Scale—for example, different scales for the same data could show a steep slope or a gentle slope, which may tell two different stories

- Interval—hour, day, month, quarter, year, and so on

- How you express the value—for example, sales could be number of widgets or dollars

- Unit of measure—percents vs. whole numbers; change on a small base could tell one story using percents, another using numbers

- Decimals—the more decimals shown, the greater degree of precision implied; sometimes rounding is more honest

Executive Summary

- In today's business environment, many of us are lost or uncertain.

- One reason why we are lost or uncertain is because we are receiving too much irrelevant data and not enough meaningful information.

- We need tools to make sense out of extreme amounts of data.

□ This book is a reference book, a sourcebook—a treasury of 101 tools to analyze and convey important business issues simply, inexpensively, and quickly yet thoroughly.

□ By having samples of 101 different types of charts available, you will be more likely to pick the correct chart in the correct situation.

□ Using these tools will help you to navigate many business situations and to gain insights faster.

□ Helping your company managers understand data will enable them to make better decisions and to perform their jobs more efficiently and effectively.

2. Tracking

This chapter contains eight charts for tracking fundamental financial concepts such as sales, cash flow, and income statements. The use of these charts and graphs will provide you with a business overview.

Tool 1. Sales and Margin vs. Time Chart

In one of my jobs, I reengineered how the finance department made presentations. The department's presentations became clearer and more concise than they had ever been. The benefit of this effort was to help the company make better sense of data, which enabled it to make better decisions and to sharpen its competitive edge.

Our before-and-after presentations of data for sales and margin vs. time provide a good example of what was done: I will first show where we started and then how we improved.

Let us begin with our initial, basic spreadsheet. (See Tool 1A.) Monthly sales and gross margin data are displayed. Gross margin is equal to sales minus the cost of goods sold. Although the numbers are accurate, there is no real analysis or information for management behind these figures.

Next, let us see how we can improve the exhibit. (See Tool 1B.) By combining the sales and gross margin dollars into quarterly data, and by adding gross margin percent, suddenly our figures have relative meaning. Gross margin percent is equal to gross margin divided by sales.

The third version (see Tool 1C) is better still, since it shows graphically, at a glance, how well the group is doing.

The fourth version (see Tool 1D) is even clearer. There is a lot going on; we have added a caption, or key takeaway box, highlighting the most significant aspect of the improvement. A key takeaway box answers the question "What does this mean?". Now that our spreadsheet has matured into a graph, the takeaway box is the key to unlocking the information hidden in the data. The key takeaway for you from these four versions: Do not force your audience to think when you should have been thinking for them.

While the key takeaway box can be used to summarize, it can also be used to suggest recommended action. In the event of poor performance, the key takeaway box should explain *where* to make changes—it can explain *how* to close the gap between actual performance and desired performance. The best business performance graphics are not just high-level variance charts. Instead, they relay which tactical changes are required.

Tool 1A

Monthly Sales and Gross Margin
Medical Systems Group

($ Millions)

	Jan	Feb	Mar	Apr	May	Jun	Jul	Aug	Sep	Oct	Nov	Dec
Net Outside Sales	6.7	9.0	11.2	7.3	9.7	12.1	7.9	10.5	13.1	8.9	11.9	14.9
Gross Margin	1.6	2.1	2.6	2.0	2.7	3.3	2.3	3.0	3.8	2.9	3.9	4.9

Tool 1B

1996 Quarterly Sales and Gross Margin
Medical Systems Group

	Q1	Q2	Q3	Q4
Net Outside Sales	26.9	29.1	31.4	35.7
Gross Margin $	6.3	8.0	9.0	11.7
Gross Margin %	23.4%	27.4%	28.8%	32.8%

Tool 1C

Sales and Margin Trend
Medical Systems Group

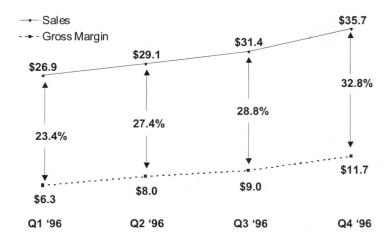

Tool 1D

Sales and Margin Trend—with Key Takeaway Box
Medical Systems Group

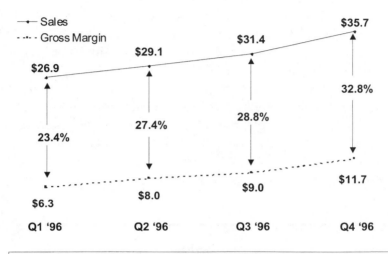

GROSS MARGIN INCREASES +9.4 PERCENTAGE POINTS, FROM 23.4% TO 32.8%

Tool 2. Stacked Column Graph

Tools 2A through 2C show ways of representing total sales by market segment. Tool 2A is a relatively typical spreadsheet, full of necessary detail. This spreadsheet contains yearly headings over columns of numbers, and one column at the left is the list of labels for the various sales subgroups. Spreadsheets are fine for certain presentations; however, this spreadsheet is cluttered. Clutter scares viewers and hides the main point. Which groups are improving year over year? From this cluttered spreadsheet, it is difficult to tell.

The data in Tool 2A could be displayed more effectively in a *stacked column graph.* Tool 2B gives an example of such a graph. Generally speaking, a stacked column graph has multiple data series stacked on top of one another, so that the top data label of each column represents the total of all the components shown in that column. These graphs are used to show how a larger entity is divided into its various components, the relative effect that each component has on the whole, and how the sizes of the components and the total change over time. A stacked column graph has multiple data series stacked on top of one another as opposed to being placed side by side. This means that in a stacked column graph, the top of the column represents the total of all the components. In Tool 2B, if each of the components represents the sales for one of the three product lines a company sells, then the top of the column represents the total sales of the company. Each data series is identified by a different shade, color, or pattern, explained in a legend.

Studies show that color matters. Color carries emotion like a ship carries cargo. Look for the feeling tones of each color and of different combinations. One of the secrets to a powerful graphic toolkit is understanding and using the impact of color. Keep in mind that this is a loose and intuitive guideline.

The stacked column graph in Tool 2C shows at a glance how each division is doing. We have shown percent to total sales for each of the divisions. For example, the tobacco company represented 44 percent of total sales in 1991, and that grows to 49 percent of sales by 1994 and 1997. The spreadsheet in Tool 2A is great for backup figures, but the graph in Tool 2C packs a wallop and gives a manager more information in less time.

Tool 2A

Sales

	1991	1992	1993	1994	1995 E	1996 E	1997 E
Cigarettes	$576,207	$123,808	$172,429	$914,573	$4,513,311	$1,013,621	$390,834
Cigars	194,461	257,671	299,981	50,305	905,602	108,022	41,772
Pipe Tobacco	135,388	423,748	659,858	142,908	270,001	3,159,701	1,868,473
Roll-Your-Own	97,009	402,071	858,681	427,247	809,222	890,052	282,651
International	129,001	316,411	620,541	444,065	315,791	104,212	40,733
Tobacco Company	1,132,066	1,523,709	2,611,490	1,979,099	6,813,927	5,275,608	2,624,462
Whiskey	90,590	810,452	988,903	300,000	605,783	175,914	400,000
Vodka	90,591	609,231	706,921	63,508	471,441	122,651	135,604
Distilled Spirits	180,181	1,419,683	1,695,824	363,508	1,077,224	298,565	535,604
Hardware	334,474	947,044	921,894	646,236	100,055	100,025	856,967
Life Insurance	77,186	733,661	908,571	161,559	543,152	761,182	267,802
Staplers	100,000	215,702	303,891	200,000	473,591	100,002	100,000
Day-Timers	100,000	350,973	412,463	100,000	125,624	785,624	100,000
International	57,288	61,039	18,978	63,508	33,038	41,959	282,044
Office Products	257,288	627,714	735,332	363,508	632,253	927,585	482,044
Subtotal	1,981,115	5,251,811	6,873,111	3,513,910	9,166,611	7,362,965	4,766,879
Specialty Business	385,932	587,956	103,586	525,067	589,978	105,579	589,165
Total Operations	2,367,047	5,839,767	6,976,697	4,038,977	9,756,589	7,468,544	5,356,044
FX/Other	1,132,953	389,823	828,631	61,023	117,321	(759,271)	(456,044)
Total Corporation	$3,500,000	$6,229,590	$7,805,328	$4,100,000	$9,873,910	$6,709,273	$4,900,000

Sales Growth %							
- Total Operations	29%	3%	19%	−42%	26%	−23%	−28%
- Total Corporation	37%	−6%	25%	−47%	27%	−32%	−27%

Tool 2B

Stacked Column Graph

Total Company Sales

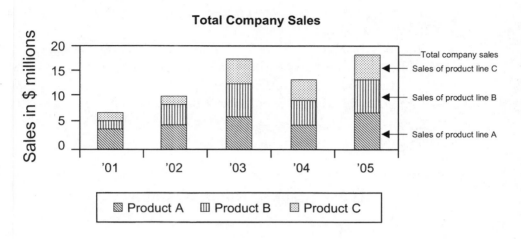

Tool 2C

Sales Summary by Market Segment

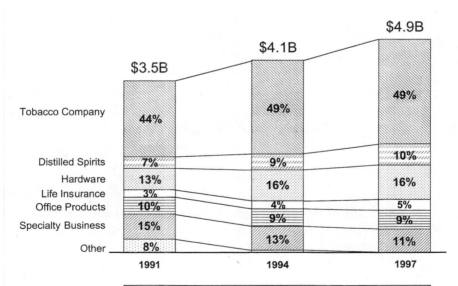

TOBACCO COMPANY INCREASES FROM 44% TO 49%
OF TOTAL COMPANY SALES

Tool 3. 100 Percent Stacked Column Graph

With a 100 percent stacked column graph, instead of plotting the actual values for each data series, the *percents* the values represent of the total of all the data series are plotted.

Tool 3 is a 100 percent stacked column graph. If the total sales for the company are $16 million and the sales for product line A in 1995 are $4 million, then 25 percent would be plotted for that product line instead of $4 million. In this type of graph, the sum of all the components, which is represented by the top line or uppermost line of the columns, is always 100 percent.

Tool 3

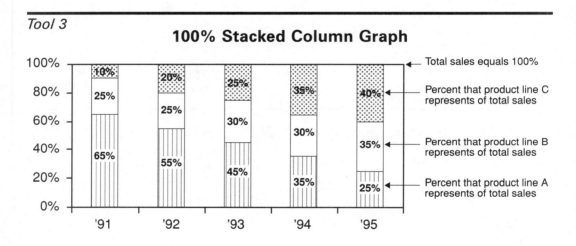

Tool 4. 100 Percent Stacked Bar Graph

Tool 4 is a 100 percent stacked bar graph. The first component of the bar for product A shows the percent that domestic sales represent of the total sales of product A. The second component shows the percent that export sales represent of the total sales of product A, and the third component shows the percent that intercompany sales represent. The sum of the components, which is represented at the far right of each bar, always equals 100 percent.

Tool 5. Wage and Salary Graph

A wage and salary graph summarizes wage and salary information in such a way that key information and relationships are graphically displayed. (See Tool 5.) Rate of pay is plotted on the vertical axis. The scale is typically graduated in values such as dollars per hour, dollars per week, or dollars per month. Wage and salary grades are shown along the horizontal axis with the lowest grade on the left. For each grade, a box is shown with the top of the box representing the maximum specified for the grade, the bottom the lowest, and a line across the middle representing the midpoint. A dot is entered for every employee in that grade. If the employee's pay is within specifications for conformity, then the dot for the employee lies within the box. If the employee's pay is above or below specifications, then the dot will lie outside the box.

The following are examples of types of information that might be learned from a wage and salary graph.

- Grades in which there are high concentrations of employees
- Distribution of employees along the entire spectrum of grades
- Whether employees are clustered in particular sections within a grade
- How many employees are above the maximum or below the minimum in each grade
- Relationships between the midpoints of each of the grades
- Spreads between maximum and minimum for the various grades
- Overlapping of pay levels from grade to grade
- Whether or not salary grades form a smooth progression

Tool 4

100% Stacked Bar Graph

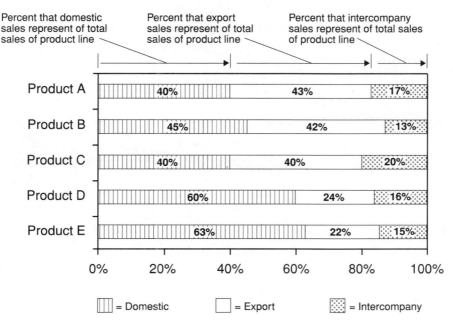

Tool 5

Wage and Salary Graph

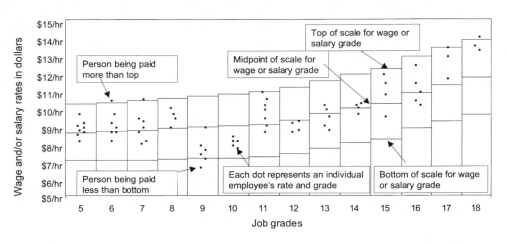

Tool 6. Cash-Flow Chart

During your career, you may have sat through a bad presentation on cash flow. I will not show you how not to present cash flow, since you can probably think of several bad examples. Tool 6 is my preferred way to present cash-flow information. In Tool 6, the columns show the three components of cash flow for each year. The three segments of each column show the impact on cash flow for changes in:

Tool 6

Cash-Flow Chart

($U.S. Millions)

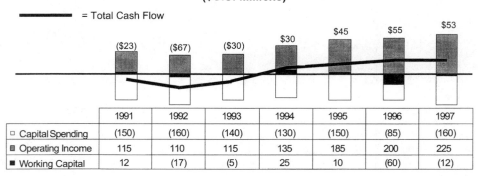

	1991	1992	1993	1994	1995	1996	1997
☐ Capital Spending	(150)	(160)	(140)	(130)	(150)	(85)	(160)
▣ Operating Income	115	110	115	135	185	200	225
■ Working Capital	12	(17)	(5)	25	10	(60)	(12)

1. Working capital

2. Operating income

3. Capital spending

The numbers and the line depict the annual cash-flow total. For example, this company had a negative cash flow of $30 million during 1993 and a positive cash flow of $30 million during 1994.

The cash-flow chart combines two different types of data graphics: the columns for the elements of cash flow and the line for the total. In this chart, a column that extends above the center line represents gross cash received (positive). A column that extends below the center line represents gross cash dispersed (negative). The line represents the net total. The columns and the line are then plotted on the same graph.

The information for gross cash outflows and the net total are plotted on the same graph so that the viewer can see both types of information at the same time. The major reason for combining the two is to improve clarity and highlight the relationships between the various elements and the total.

Tool 7. Waterfall Chart

The waterfall chart, also called a cascade chart, is an excellent way to illustrate quantitative flows, or how you get from number A to number B. This chart (see Tool 7A) shows the total sales for SBJ Technology, Inc. The columns show components of sales, broken down into stainless steel, super alloys, tool steel, and structural ceramics. The last column is total sales. A waterfall chart can also be used to depict a simplified income statement, starting with sales on the left and ending with net income on the right, and showing the various items that lead from one to the other. The starting point (sales in the example) is always a column that begins at zero. (See Tool 7B.)

Waterfalls can depict static data (balance sheets, income statements) or active data (time series data, cash flows). You can mix negative and positive items. For example, you can tell the following story with a waterfall chart: We started with $150 million in net widget sales and ended with $6 million in net income. The waterfall chart segregates positive items (net widget sales, interest income, and gain on sale of gadgets subsidiary) and negative items (operating expenses and taxes) to show, say, where value is created or lost.

Whatever data you use, the waterfall chart is a versatile way to convey a lot of information in a clear and concise manner.

Tool 7A

Waterfall Chart—Sales
SBJ Technology, Inc.

Total Sales, 1999

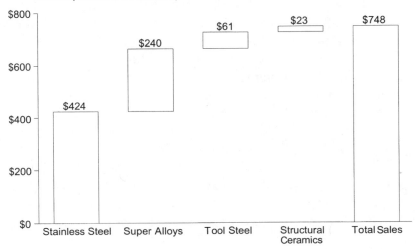

Tool 7B

Waterfall Chart—Income
Shaskan Widget Corporation

Income Statement

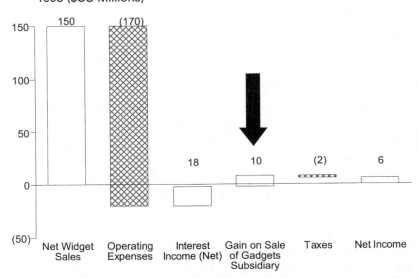

We would have posted a loss in 1998 without gain on sale of our gadgets division.

Tool 8. Thermometer Chart

Most people are familiar with thermometer charts; they resemble a thermometer, showing progress toward a goal. Usually they are percentage calibrated. In our example (see Tool 8), the worksheet tracks daily progress toward a goal of achieving 1,000 new customers in a 15-day period. This example shows that as of Day 6 of our campaign, we have 626 new customers; the chart shows visually that we are now 63 percent of the way toward our goal.

How to re-create the thermometer chart using Excel

We assign cell B18 (1,000 in Tool 8) as the goal value. Cell B19 (626 in Tool 8) contains the formula =SUM(B2:B16), and cell B21 (63% in Tool 8) contains =B19/B18, a formula for calculating the percentage of the goal attained. As new data is entered in column B, the formulas display the current results.

To create the chart, enter the formulas listed above, along with the figure's sample data; then select cell B21, and click the Chart Wizard button.

Tool 8

Thermometer Chart

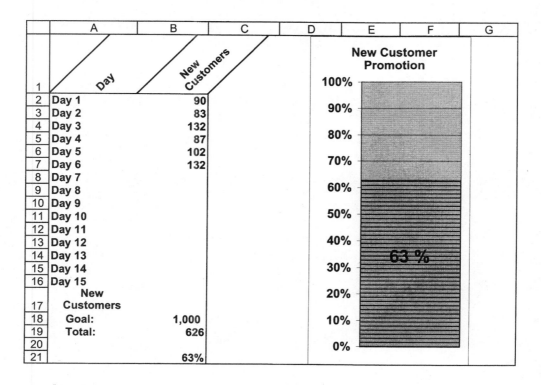

Notice the blank row preceding cell B21. If you fail to include this blank row, Excel will use the entire data block—not just the single cell—to construct the chart. Since cell B21 is isolated from the other data, the Chart Wizard uses only the single cell. In step 1 of the Chart Wizard dialog, specify a Column chart and a Cluster Column subtype (the first choice). Click Next twice, and then in step 2 make additional adjustments: Add a Chart Title (Title tab), dump the Category (X) axis (Axes tab), delete the legend (Legend tab), and specify Show value (Data Labels tab). Click Finish to view the chart.

The chart needs further customization. To display the Format Data Series dialog, double-click the column. Click the Options tab, and set the Gap width to 0. (This setting instructs the column to occupy the entire width of the plot area.) To change the pattern used in the column, click the Patterns tab and make your selection. The example shown here uses a gradient fill effect. Next, double-click the vertical axis to bring up the Format Axis dialog. In the Scale tab of the Format Axis dialog, set Minimum to 0 and Maximum to 1.

3. Variances and Comparisons

This chapter covers eight techniques to display systems of interrelated values so that deviations, groupings, and the relative sizes of more than one element can be seen at the same time.

Tool 9. Deviation Graph

A deviation graph displays the differences between a data series and some known reference, such as a budget, industry standard, or prior year's results. Deviation graphs often are used to display performance to standard cost, performance to schedule, performance to cost reduction goals, or performance to departmental budgets. They are useful whenever there is a need to display and analyze variances, since a variance is any deviation from a predetermined benchmark.

In this example, budget is used as the reference. Tool 9A shows actual data. Tool 9B is a deviation graph, since it shows the difference (deviation) between actual profit and budget. When the actual profit value is above budget, the difference is denoted as positive; when profit is below budget, the difference appears as negative; and when profit is equal to budget, the deviation is zero.

Tool 9A

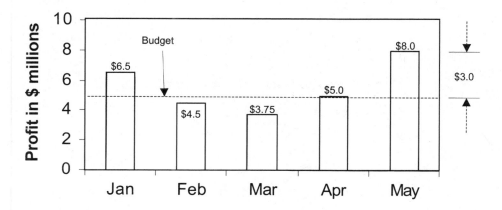

Actual Data

Tool 9B

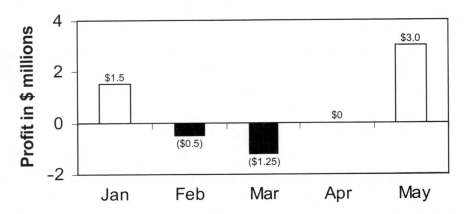

Deviation Graph

Tool 10. Horizontal Bar Graph

As we have seen with Tool 4 (100 percent stacked bar graph), certain concepts lend themselves well to a horizontal orientation. And as we have seen with Tool 9B (deviation graph), observing deviations can be crucial to effective management decisions. That is why I am introducing the horizontal bar graph here as another effective way to display the story behind the numbers.

In Tool 10, the horizontal bar graph, column B contains the budget target; column C, the actual spending; and column D, the monthly budget variance percent. Each bar represents variance percent, and the values are shown on the adjacent data table in column D. The major purpose of the graphical portion is to orient the viewer visually to the relative sizes of the budget vs. actual variances. Here the formulas in columns E and G (where we see the small black squares) depict monthly budget variances. This chart can easily be re-created in Excel.

How to re-create the horizontal bar graph using Excel

Enter the data shown in columns A through D, and then enter the following formulas.

```
E2 = If(D2<0,rept("n",-round(D2*100,0)),"")
F2 = A2
G2 = If(D2>0,rept("n",-round(D2*-100,0)),"")
```

Assign the Wingdings font to cells E2 and G2, and then copy the formulas down the columns to accommodate all the data. Center-align the

Tool 10

Horizontal Bar Graph

	A	B	C	D	E	F	G
1		Budget	Actual	Variance %	Below budget		Above budget
2	Jan	300	311	3.7%		Jan	■■■■
3	Feb	300	298	-0.7%	■	Feb	
4	Mar	300	305	1.7%		Mar	■■
5	Apr	350	351	0.3%		Apr	
6	May	350	402	14.9%		May	■■■■■■■■■■■■■■
7	Jun	350	409	16.9%		Jun	■■■■■■■■■■■■■■■■
8	Jul	500	421	-15.8%	■■■■■■■■■■■■■■■■	Jul	
9	Aug	500	454	-9.2%	■■■■■■■■■	Aug	
10	Sep	500	474	-5.2%	■■■■■	Sep	
11	Oct	500	521	4.2%		Oct	■■■■
12	Nov	500	476	-4.8%	■■■■■	Nov	
13	Dec	500	487	-2.6%	■■■	Dec	

text in column F, right-align column E, left-align column G, and adjust any other formatting as you like. Depending on the numerical range of your data, you may need to change the scaling. Experiment by replacing the "100" value in the formulas. You can, of course, substitute any character you like for the *n* in the formulas to produce a different character in the chart.

This technique also works in 1-2-3 and Quattro Pro. Just use the @repeat function in place of Excel's Rept function, and insert an "at" sign (@) before the If and Round functions.

Tool 11. Side-by-Side Column Graph

A side-by-side column graph is a column graph with two or more data series plotted side-by-side for comparison purposes. (See Tool 11.) The columns for a given data series are always in the same position in each group throughout a given graph. In our example, each Plant makes the same product, and the vertical axis displays percent of total cost. The information for Plant A is always at the left of each group, followed by the information for Plants B and C. Each data series is a different color, shade, or pattern.

Tool 12. Side-by-Side Bar Graph

A side-by-side bar graph is a bar graph with two or more data series plotted side-by-side for comparison purposes. (See Tool 12.) The bars for a given data series are always in the same position in each group throughout a given graph. In our example, the information for Plant A is always at the top of each group, followed by the information for Plants B and C. Each data series is a different color, shade, or pattern.

Tool 11

Side-by-Side Column Graph
U.S. Tool Company

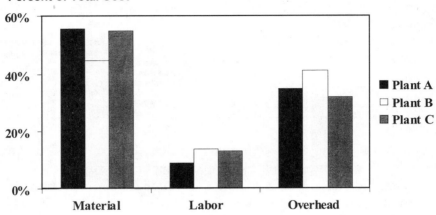

Percent of Total Cost

Legend: Plant A, Plant B, Plant C

Categories: Material, Labor, Overhead

Tool 12

Side-by-Side Bar Graph
U.S. Tool Company

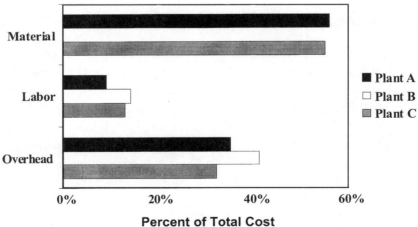

Percent of Total Cost

Legend: Plant A, Plant B, Plant C

Categories: Material, Labor, Overhead

Tool 13. Histogram/Pareto Chart

A Pareto chart is a special type of histogram (column graph) that visually compares numerical data. It displays, from greatest to least, the frequency of occurrences, such as types of problems, costs of problems, problems by product, or various other measurements to help identify the most significant problems that need to be corrected. It is a column graph used to arrange information in such a way that process improvements can be prioritized.

Pareto charts are named after Vilfredo Pareto, an Italian sociologist and economist, who invented this method of information presentation toward the end of the nineteenth century. The chart is a form of histogram or column graph (see Tool 13A) with the columns arranged in decreasing order from left to right along the X axis (see Tool 13B). The fundamental idea behind the use of Pareto diagrams for quality improvement is that the first few (as presented on the diagram) contributing causes to a problem usually account for the majority of the result. Thus, targeting these "major causes" for elimination results in the most cost-effective scheme.

Example: Based on the Pareto chart (see Tool 13B), a team might want to attack the problem of "late shipments" first. By reducing late shipments, the team can reduce the majority of shipping problems. This is an example of the "80/20 rule of thumb" (often roughly 80 percent of an overall effect can be traced back to about 20 percent of all contributing factors). The Pareto chart works best solving problems that revolve around frequency. Bear in mind that if shipping the wrong item is a more costly problem than a late shipment would be, then the Pareto chart by itself would not be the best tool for this analysis.

Use Pareto charts to:

• Allocate problem-solving resources.

• Track the effectiveness of problem-solving efforts.

• Measure the impact of solutions.

Tool 13A

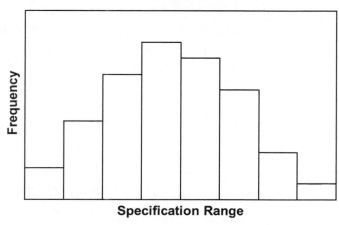

Histogram

Frequency

Specification Range

Tool 13B

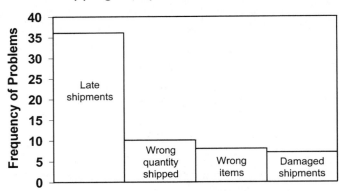

Pareto Chart
Shipping Complaints for Month of July

Tool 14. Radar Chart

A radar chart is sometimes referred to as a star or spider chart. A radar chart is a circular chart used primarily as a comparative tool. For example, on a radar chart, the nutritional content of two different foods might be compared based on the percentage of recommended daily allowances of five different vitamins that each contains. Tool 14 is an illustration of such a comparison. In this example, both Food A and Food B have an equal amount of riboflavin (30 percent) per serving. However, Food A has 60 percent of the recommended daily allowance of niacin, while Food B has only 40 percent. Comparing these five factors, Food A would be considered to have higher overall nutritional value than Food B.

Radar charts can be interpreted either by reading the actual values on the axes or by comparing the areas enclosed by the polygons formed by the data points. In this example, the larger the polygon, the "better" the food. In general, whether a smaller or larger polygon is "better" depends on whether the more favorable values are nearer the center or the circumference. Whichever convention is selected, it is used consistently throughout the chart. When one polygon is distinctly larger than the other, the interpretation of the graph is straightforward. In other cases, where the lines intersect one another, the interpretation is more difficult and many times involves a judgment factor.

In radar graphs, each category axis represents a different variable. Other than readability, there is no limitation on the number of variables that can be included in a single graph. Whatever the number of variables, they are distributed equally around the 360 degrees of the circle.

Tool 14

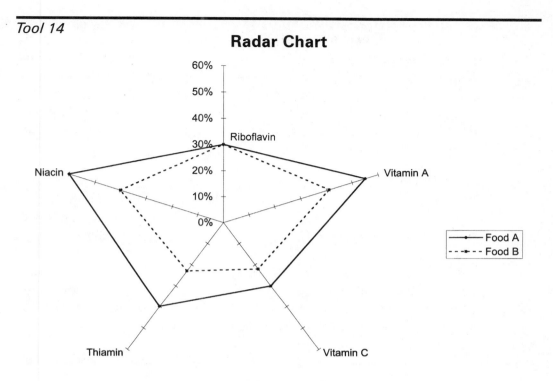

Tool 15. Marimekko Graph

A Marimekko graph is the graph of choice for many strategic consulting firms throughout the world. With the ability to communicate complex data and concepts in a single two-dimensional graph, Marimekkos are used extensively in corporate boardrooms.

The graphs are so named because they bear a striking resemblance to the carpets and tapestries produced by the Finnish manufacturer of the same name. They also look like mosaics. A Marimekko graph is made up of a series of interspersed 100 percent stacked column graphs and 100 percent stacked bar graphs. Its major function is to display a system of interrelated values in such a way that groupings and relative sizes of the many elements can be seen at the same time. The steps involved in constructing such a graph are used to describe the nature of the graph. Our example is of a manufacturing company; however, a Marimekko graph can be used for many different applications.

Step one

This example (see Tool 15A) analyzes how a company's overall sales dollars are allocated. Therefore, the full length of the Marimekko graph, which is a 100 percent stacked bar graph (included as Tool 4), represents total sales dollars ($50 million). The components of the bar graph are the five major categories for which the sales dollars are used (Cost to manufacture, Research and Development (R&D), Marketing, Administrative (Admin.), and Profit). A zero-to-100 percent scale is displayed at the bottom of the graph, so the viewer can determine the percent that each component represents of total sales. For example, 50 percent ($25 million) of total sales was used to manufacture the product. Actual sales values may or may not be shown at the top.

Step two

A more detailed Marimekko graph (see Tool 15B) shows how the dollars are allocated within each major component. Each individual segment of the horizontal bar graph becomes its own 100 percent stacked column graph (included as Tool 3). For example, in the cost to manufacture, it can be seen that there are three major types of costs: material, labor, and overhead. It also can be seen that material alone accounts for 55 percent of the total manufacturing costs. Similar stacked column graphs are clearly outlined for R&D, Marketing, and Administrative.

Tool 15A

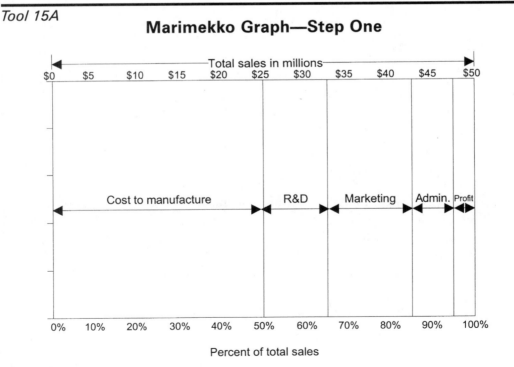

Marimekko Graph—Step One

Tool 15B

Marimekko Graph—Step Two

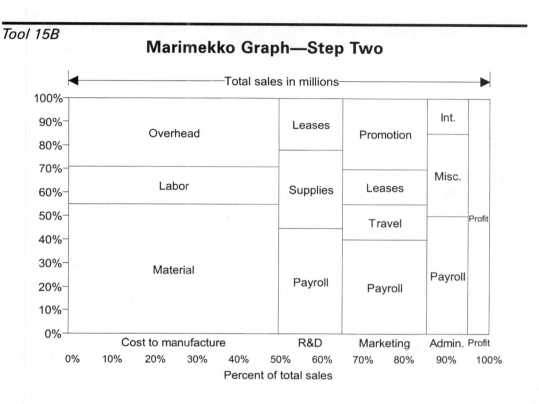

Tool 16. Marimekko Column Graph

A Marimekko column graph is a variation of a column graph in which the widths of the columns have significance. It is also referred to as an area column graph. In a Marimekko column graph, the widths of the columns are proportional to some measure or characteristic of the data elements represented by the columns. For example, if the columns are displaying the profitability of various product lines, then the width of the columns might indicate what percentage of total sales the various products represent. Column widths can be displayed in terms of percents or units. In our example, the total value is noted in the key takeaway box. (See Tool 16.) If the values along the horizontal axis are cumulative, then the columns are generally joined so that there are no unaccounted-for spaces. For example, if all of the columns add up to a certain value, such as 100 percent, or if the sum of all of the values along the horizontal axis is important, no spaces are left between columns. Spaces can be used between the columns if the values are not cumulative, if the total is unimportant, or if a scale is part of the legend. Sometimes the widths and heights of the columns are related such that as one varies, the other varies also. For example, in a histogram, if the width of a column is increased to encompass a broader class interval, then the height of the column is adjusted accordingly. In these cases, the area of the column is a more accurate measure than either the height or the width. The concept of using the width of columns to convey additional information is generally applied only to simple and stacked column graphs, including 100 percent stacked graphs (included as Tools 3 and 4).

Tool 16

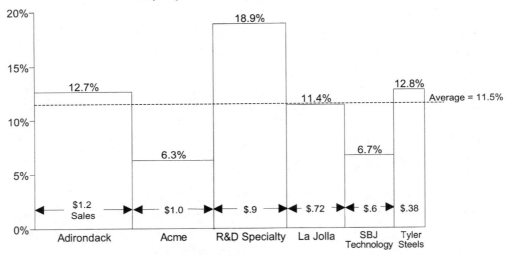

Marimekko Column Graph
U.S. Specialty Steel Market

Company Profitability per Dollar of Sales

AVERAGE RETURN ON SALES 11.5 PERCENT
ON TOTAL SALES OF $4.8 MILLION

4. Trends and Change

This chapter presents 13 ways to depict trends, change, and quality control procedures.

Tool 17. Rolling Total Curve

On a rolling total curve, the value plotted for each interval is equal to the sum of the incremental value for that interval, plus the values for a fixed number of preceding intervals. For instance, if one is preparing an annual rolling total curve, then the value plotted in May would be equal to the actual value for May plus the combined values for the preceding 11 months, making it a 12-month or annual total. In June, the plotted value would be equal to the actual value for June plus the combined values for the preceding 11 months. Any number of intervals can be used in the total. Whatever number of intervals is selected, that same number is used consistently throughout the entire graph. Actual values often are plotted on the same graph with the rolling total. (See Tool 17.) The purpose of rolling total curves is to smooth out the interval-to-interval fluctuations somewhat and provide a graphic indication of the overall trend of the data. In our example, the rolling total indicates a general decline in the annualized totals until about the first quarter of the second year, at which time it tends to increase. That trend is not obvious in the monthly actual values.

Tool 18. Moving Average Graph

A moving average graph is a method used to smooth the curve of a data series and make general trends more visible. It is sometimes referred to as a rolling average graph or trend line graph. The method involves generating a second curve of a data series with the short-term peaks and valleys smoothed out. (See Tool 18A.) The degree to which the peaks and valleys are smoothed depends on the number of intervals used. Each point on a moving average curve is generally calculated by averaging the value for the current period plus a fixed number of prior periods. Each time the value for a new period is added, the value for the oldest period in the previous calculation is dropped. For example, if monthly sales data were being tracked, a four-month period might be used for the average. Thus, in April, the values for January, February, March, and April would

Rolling Total Curve

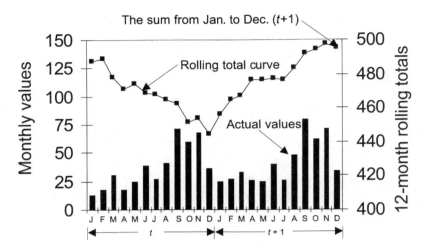

be averaged and that point plotted. In May, the values for February, March, April, and May would be averaged and that point plotted. The number of prior time periods included in the average can vary significantly. Three to 200 are commonly used, though there is no limit on the number of periods that can be included in the averaging process.

To re-create Tool 18A, we have included the moving average graph worksheet as Tool 18B. On the worksheet, the "Actual" column contains the data for the first curve that is on the graph. The "4-Month Average" column contains the data for the second (bold) curve. The four-month average for August (39) is the average of the data for May, June, July, and August.

Tool 19. Index Graph

An index graph is a graph that displays values in terms of a percent, fraction, or ratio of some reference value. For example, a graph might show the value of a house each year as a percent of its purchase price. (See Tool 19A.) If a house purchased in 1995 for $200,000 is worth $400,000 in 1997, then it would be plotted at the 200 percent value. The 200 percent is considered an index value. If an index includes the values of several factors, as the consumer price index does (combining data on food, housing, fuel, etc.), it is sometimes referred to as a composite index.

There is no limit to the number of data series that can be plotted on a single graph. Sometimes only the beginning and ending or current index values are plotted. This format sometimes makes the data stand out more crisply, particularly when there are many values. Many times this format is called a comparative index graph. (See Tool 19B.)

A second vertical axis on the right-hand side can be a helpful addition to make an index graph even clearer.

Tool 20. Seasonal Graph

Some organizations experience major seasonal fluctuations in certain aspects of their operations. For example, with some companies, the effects of the yearly cycle may make it difficult to spot trends in monthly sales. For these companies, most of their sales might occur over the year-end holiday season. With others, fluctuations in sales, expenses, and manpower might occur during the summer or be associated with some annual occurrence, such as the sale of roses on Valentine's Day. When data is plotted for these organizations, sometimes there are such significant fluctuations that it is difficult to spot trends or note unusual deviations. To address this problem, specialized index graphs have been developed that display mathematically adjusted data that removes or smooths out seasonal fluctuations. In this way the effects of the seasonal cycle can be negated, and the trends and variations in the data can be observed more easily. This is accomplished by using historical data to develop a series of seasonal indexes. Index numbers are normally calculated for each month of the year; however, other periods, such as weeks or quarters, might be used. On the assumption that seasonal fluctuations

Tool 18A

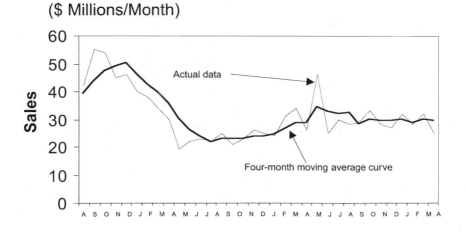

Moving Average Graph

($ Millions/Month)

Moving Average Graph Worksheet

	Actual	4-Month Average
May	37.0	
June	38.0	
July	39.0	
August	42.0	39.0
September	55.0	43.5
October	54.0	47.5
November	45.0	49.0
December	46.0	50.0
January	40.0	46.3
February	38.0	42.3
March	34.0	39.5
April	30.0	35.5
May	19.0	30.3
June	22.0	26.3
July	23.0	23.5
August	22.0	21.5
September	25.0	23.0
October	21.0	22.8
November	23.0	22.8
December	26.0	23.8
January	25.0	23.8
February	24.0	24.5
March	31.0	26.5
April	34.0	28.5
May	26.0	28.8
June	46.0	34.3
July	25.0	32.8
August	30.0	31.8
September	28.0	32.3
October	29.0	28.0
November	33.0	30.0
December	28.0	29.5
January	27.0	29.3
February	32.0	30.0
March	28.0	28.8
April	32.0	29.8
May	25.0	29.3

Tool 19A

Index Graph

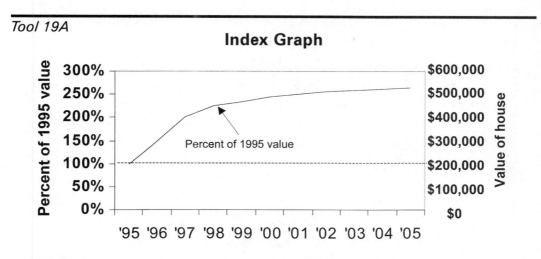

Tool 19B

Comparative Index Graph

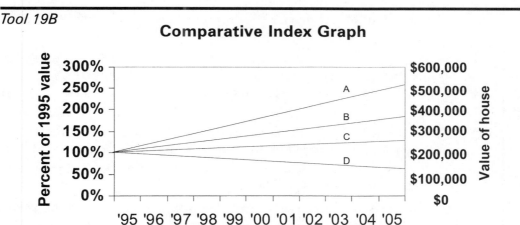

are the same from year to year, the seasonal indexes are then applied to past data or future projections and the resulting values plotted.

Tool 20A is a graph in which data has not been adjusted for seasonal variations. Tool 20B shows the same data except that it has been seasonally adjusted and a regression line added. After the data has been seasonally adjusted, it is easier to draw meaningful conclusions. The downward trend of the data in this example has now become clearer. In addition, certain unusual or unexplained variations stand out—for example, the greater-than-normal swing of values in 1996 and the fact that the values in the second half of 1996 were higher than the first half, which is contrary to the pattern in 1997.

Tool 20C contains the actual data for these graphs. To help you re-create these graphs, here is an explanation of the columns on the worksheet.

- YEAR: The first column contains the year, 1996 through 1999.

- PERIOD: This counts the number of periods for which we have data. The worksheet contains data for 48 periods.

- MONTH: The third column contains the month.

- ACTUAL SALES: Sales are actual data for January 1996 through December 1999. The data in this column has *not* been adjusted for seasonal variations. This is the data that has been graphed in Tool 20A.

- MOVING AVERAGE: Moving average is a rolling 12-month average for "Actual Sales." The first moving average (for July 1996) is the average of the data for January through December 1996.

- RATIO: This is the actual sales divided by the moving average.

- SUM OF RATIOS: This adds together all the ratios for a particular month. For example, the number in the "February" row, 4.33, is the sum of the ratios for February 1997 + February 1998 + February 1999 (1.57 + 1.37 + 1.39 = 4.33).

- NUMBER OF RATIOS: This is the number of ratios for each particular month. This worksheet calculates three ratios for each month, except for July. (There are four ratios for July, one each for the years 1996 through 1999.)

- AVERAGE RATIO: This is the average of the ratios (i.e., the "Sum of Ratios" column divided by the "Number of Ratios" column). This calculation is required because the ratios for the corresponding periods in the four years differ significantly.

- SEASONAL INDEX: The average ratios have been normalized, which means that they have been adjusted proportionally so that they now add up to 12.

- ADJUSTED SALES: This is the "Actual Sales" for each period divided by the "Seasonal Index" for each period. This column contains the seasonally adjusted data. It is the data that has been graphed in Tool 20B.

Tool 21. Composite/Decomposite Graphs

Sometimes the data from which a graph is plotted is made up of several components. When analyzing sales, for example, a five-year sales curve might include the continuing growth of the company, the cyclical fluctuations of the economy, seasonal fluctuations, and random fluctuations due to such things as special promotions. To reveal the long-term trend, the systematic and irregular components of total sales can be identified and graphed separately. When the data for those components is separated out and plotted individually, the resulting set of graphs are called de-composite graphs. Tool 21A is a composite graph. Tool 21B displays decomposite graphs, plotting the four components that make up the composite graph.

As an example, a composite graph might contain five years of sales data for a company that sells lawn mowers. The decomposite graphs for this company might break total sales down into four separate components.

- GENERAL TREND: This reveals that the long-term trend for lawn mower sales is favorable.

- CYCLICAL FLUCTUATIONS: Cyclical fluctuations might be due to a

Tool 20A

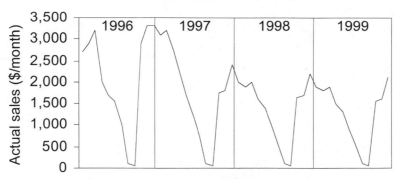

Seasonal Graph—Unadjusted

Graph in which data has not been adjusted for seasonal variations

Tool 20B

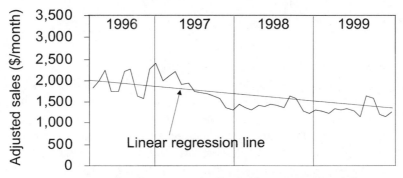

Seasonal Graph—Adjusted

Graph with curve adjusted for seasonal variations

Tool 20C

Seasonal Graph Worksheet

Year	Period	Month	Actual Sales	Moving Average	Ratio	Sum of Ratios	Number of Ratios	Average Ratio	Seasonal Index	Adjusted Sales
1996	1	Jan	2,700.0			4.40	3	1.47	1.47	1,834.77
	2	Feb	2,900.0			4.33	3	1.44	1.45	1,999.33
	3	Mar	3,200.0			4.24	3	1.41	1.42	2,255.70
	4	Apr	2,000.0			3.40	3	1.13	1.14	1,760.13
	5	May	1,700.0			2.90	3	0.97	0.97	1,748.81
	6	Jun	1,550.0			2.10	3	0.70	0.70	2,209.15
	7	Jul	1,000.0	2,058.33	0.49	1.75	4	0.44	0.44	2,270.77
	8	Aug	100.0	2,091.67	0.05	0.18	3	0.06	0.06	1,634.73
	9	Sep	50.0	2,116.67	0.02	0.09	3	0.03	0.03	1,598.55
	10	Oct	2,900.0	2,075.00	1.40	3.82	3	1.27	1.28	2,268.94
	11	Nov	3,300.0	2,091.67	1.58	4.12	3	1.37	1.38	2,392.91
	12	Dec	3,300.0	2,091.67	1.58	4.96	3	1.65	1.66	1,988.81
1997	13	Jan	3,100.0	2,062.50	1.50		Sum	11.95	12.00	2,106.59
	14	Feb	3,200.0	2,041.67	1.57					2,206.16
	15	Mar	2,700.0	2,041.67	1.32					1,903.24
	16	Apr	2,200.0	2,041.67	1.08					1,936.14
	17	May	1,700.0	1,945.83	0.87					1,748.81
	18	Jun	1,200.0	1,820.83	0.66					1,710.31
	19	Jul	750.0	1,745.83	0.43					1,703.08
	20	Aug	100.0	1,654.17	0.06					1,634.73
	21	Sep	50.0	1,545.83	0.03					1,598.55
	22	Oct	1,750.0	1,487.50	1.18					1,369.19
	23	Nov	1,800.0	1,437.50	1.25					1,305.22
	24	Dec	2,400.0	1,412.50	1.70					1,446.41
1998	25	Jan	2,000.0	1,395.83	1.43					1,359.09
	26	Feb	1,900.0	1,383.33	1.37					1,309.91
	27	Mar	2,000.0	1,383.33	1.45					1,409.81
	28	Apr	1,600.0	1,383.33	1.16					1,408.11
	29	May	1,400.0	1,375.00	1.02					1,440.20
	30	Jun	1,000.0	1,366.67	0.73					1,425.26
	31	Jul	600.0	1,350.00	0.44					1,362.46
	32	Aug	100.0	1,341.67	0.07					1,634.73
	33	Sep	50.0	1,333.33	0.04					1,598.55
	34	Oct	1,650.0	1,325.00	1.25					1,290.95
	35	Nov	1,700.0	1,316.67	1.29					1,232.71
	36	Dec	2,200.0	1,308.33	1.68					1,325.87
1999	37	Jan	1,900.0	1,300.00	1.46					1,291.14
	38	Feb	1,800.0	1,291.67	1.39					1,240.96
	39	Mar	1,900.0	1,291.67	1.47					1,339.32
	40	Apr	1,500.0	1,291.67	1.16					1,320.10
	41	May	1,300.0	1,283.33	1.01					1,337.33
	42	Jun	900.0	1,275.00	0.71					1,282.73
	43	Jul	500.0	1,266.67	0.39					1,135.39
	44	Aug	100.0							1,634.73
	45	Sep	50.0							1,598.55
	46	Oct	1,550.0							1,212.71
	47	Nov	1,600.0							1,160.20
	48	Dec	2,100.0							1,265.61

host of factors. Economic cycles might affect sales, since fewer lawn mowers are sold during a recession. Cyclical weather patterns might affect sales, because more rainfall means longer grass, which in turn means more lawn mower sales. Cyclical changes in commodity prices might affect lawn mower sales, because fewer lawn mowers are sold when they are more expensive, and the price of lawn mowers is directly related to the cost of aluminum and other raw material inputs.

- SEASONAL FLUCTUATIONS: Perform this type of sales analysis by calculating a seasonal index. (See Tool 20.) For lawn mowers, this would show that more are sold in summer months and fewer are sold during winter months.

- RANDOM FLUCTUATIONS: This equals the variation in sales that is not explained by any of the other three factors already mentioned. It represents the difference between "Total sales" and the sum of "General trend" + "Cyclical fluctuations" + "Seasonal fluctuations."

Tool 22. Silhouette Graph

A silhouette graph is a series of small, related graphs stacked on top of one another to provide an overview of a subject, project, program, or financial situation over time. A silhouette graph uses shading or patterns to fill the areas between the data series and the reference line. Each of the multiple graphs uses the same horizontal scale located on the lower axis, but each has its own vertical axis. However, units of measure on the vertical axes of the various graphs may or may not be the same. In our example, the units of measure are the same, and the intervals on the scales are the same so that the silhouettes can be compared.

Tool 22 is a silhouette graph used as an overview of costs on a major project. Similar graphs can be used to show other aspects, such as manpower, equipment, and cash requirements. Since they contain many series of data, silhouette graphs are used primarily for overview and planning purposes. A worksheet is more suitable for tracking purposes, since it provides more detail to analyze the data.

Tool 21A

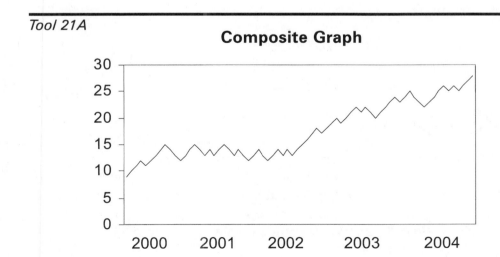

Composite Graph

Tool 21B

Decomposite Graphs

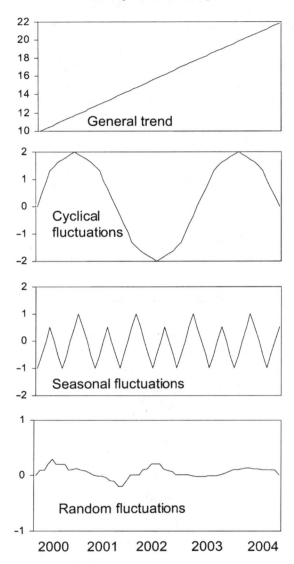

Tool 23. Channel Chart

In a channel chart, a channel, or envelope, is generated when a pair of lines is drawn at the top and bottom of a series of data points. The lines may be drawn freehand or plotted based on some mathematical procedure. They may or may not encompass all of the data points. The purpose of the lines is to approximate the boundaries of the data to assist the viewer in estimating the general shape and trend of a group of data points.

One of the principles of technical analysis of stocks is that prices tend to move in trends. Tool 23 has two lines connecting the tops and bottoms, creating a "channel" to delineate the uptrend. In the charting that is done for the technical analysis of stocks, a price channel contains prices throughout a trend. One of the first things the chartist looks for is a trend. As John Magee wrote in the bible of charting, *Technical Analysis of Stock Trends*, "Prices move in trends, and trends tend to continue until something happens to change the supply-demand balance."

Tool 22

Silhouette Graph

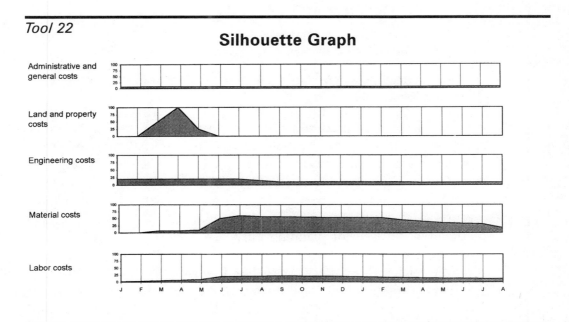

Tool 23

Channel Chart

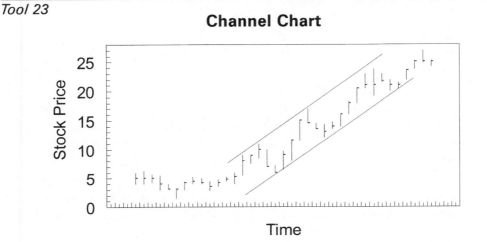

Time

Tool 24. Product Life Cycle Graph

The sales of many products follow a typical life cycle over the length of their existence. That cycle generally consists of five major phases.

1. The period during which the product is being developed and there are no sales

2. The period during which the product is introduced to the market and sales first begin to be realized

3. The growth phase, when the product has been accepted and sales rise at their fastest rate

4. The leveling stage, as the product matures and other competitive products are introduced

5. The decline, when sales decline as the need or desire for the product decreases

Tool 24A is often used to graphically illustrate the product life cycle. Although the general shape of the sales curve is typically the same from product to product, the length of time required for each phase may vary significantly. Unlike the idealized graph, the length of the various phases is not always the same. Product life cycle graphs are used frequently for conceptual and planning purposes as well as for actual tracking of a product. When used for planning purposes, curves such as profit, cash flow, and capital expenditures often are added to show how these measures are related to the sales cycle. (See Tool 24B.) In Tool 24C, multiple sales curves are drawn to illustrate how the repeated introduction of new products can help to offset the revenue decline of older products and thus assist in maintaining relatively smooth sales.

Tool 25. Bathtub Curve

A bathtub curve is so named because of its distinctive shape: higher values at the beginning and end with relatively constant, lower values in between. (See Tool 25.) Time is plotted on the horizontal axis, with zero at the left and time increasing to the right. A bathtub curve is sometimes used to illustrate how the failure rates of certain products change with time. When used in this way, time zero is considered the time at which each product first goes into service. Periods of time, usually uniform, are measured along the horizontal axis. The percent of failures for each period of time is plotted on the vertical axis. The higher failure rates at

Tool 24A

Product Life Cycle Graph—Tracking

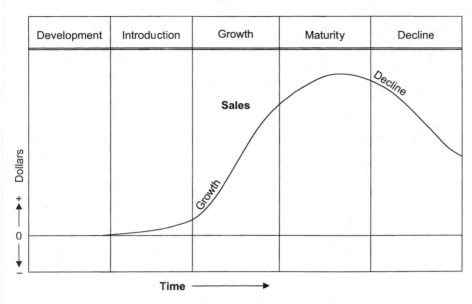

Tool 24B

Product Life Cycle Graph—Planning

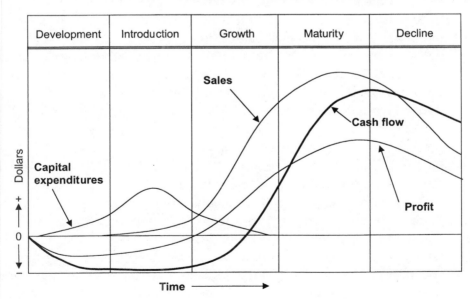

Tool 24C

Product Life Cycle Graph—Replenishing

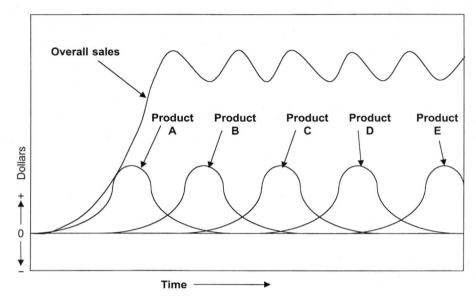

the left of the graph occur shortly after the product is put into service. After these early failures, the rate typically drops to a low level, where it stays for the rest of the product's expected life. If the product continues to be used beyond its normal life expectancy, then the failure rate begins to rise again as the product wears out.

As an example, consider any small electric household appliance, such as a blender. A manufacturer might make thousands of blenders in any given year, and it might have a warranty policy that allows consumers to return blenders when they fail. The bathtub curve depicts when these blenders fail relative to when they are sold. Since no manufacturing process is perfect, when a blender is first taken out of the box and plugged in, it will either work or not work. The brand-new blenders that do not work are depicted as "early failures" on the left of the graph. If a blender does work, then it probably will continue to work for some period of time. In other words, there will be a period of time when the blenders have a low failure rate. This is depicted as the flat middle section of the graph. Eventually, of course, the blenders will wear out and fail. The failures that come with time are depicted on the right of the graph. By analyzing the shape of the bathtub curve, the appliance manufacturer can make better decisions about how to improve the manufacturing process and how long the blender should be covered under its warranty return policy.

Tool 25

Bathtub Curve

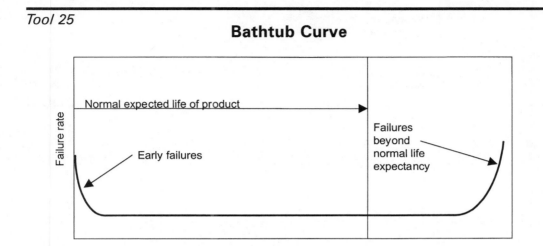

Tool 26. Conformance Graph

The conformance graph comes from the Japanese statistician Genichi Taguchi, who emphasizes analysis of the distribution of variations from specifications. His research indicates that it is better to produce parts whose distribution is centered on the target specification than parts whose distribution is not centered on the target, even if the latter distribution is tighter. If one part is at the upper end of its specified limits and the other at the lower, then the *combined* variation from specifications may prevent the two parts from fitting together properly. Taguchi's approach to quality conformance is to make sure the distribution of each part is such that the number of cases of tolerance stack-up is minimized. The approach is useful in cases where multiple parts are being made for assembly into products. It can be applied to either manufactured or purchased parts.

Note: A number of U.S. consulting firms use similar types of analysis under the name *statistical process control*.

There are three steps to follow when using a conformance graph.

1. Draw a large sample of parts from the production line (or from those supplied by vendors)—at least 30 and preferably more. Measure them. Use a separate graph for each of the specified measurements if there are more than one.

2. Analyze the distribution of measurements; count the number of parts with like measurements and create a table, as in the following example for a sample of 70 parts.

Length (millimeters)	Number
4.10	1
4.11	1
4.12	3
4.13	5
4.14	11
4.15	21
4.16	14
4.17	6
4.18	4
4.19	3
4.20	1

3. Plot this data on a graph with the horizontal axis for the measurements and the vertical axis for frequency. Draw a curve to represent the target specification.

In Tool 26, the data from the table is plotted on a column chart and the target is indicated. The sample is not centered on its target, and the graph therefore indicates a potential problem that should be addressed before further production. The graph also shows that the frequency for one part is greater than 20 (i.e., the part with a length of 4.15 has a frequency of 21); an action limit has been exceeded, which is a further indication of trouble. But according to Taguchi's theory, this problem is less important than the fact that the center of the distribution is off target.

Use a conformance graph to:

• Analyze the conformance of parts to their specifications.

• Train employees in the principles of statistical process control in manufacturing, logistics, and other settings.

Tool 27. Control Chart

A control chart operates on the principle of monitoring some quantifiable characteristic of a repetitive process or operation. Actual values of those characteristics are recorded and plotted on a time series graph. A time series graph is a simple point or line graph that has a time series scale on the horizontal axis, with time progressing from left to right. A control chart characterizes the behavior of the time series. Occasionally one will encounter a time series that is predictable, consistent, and stable over time. More commonly, time series are unpredictable and inconsistent and

Tool 26

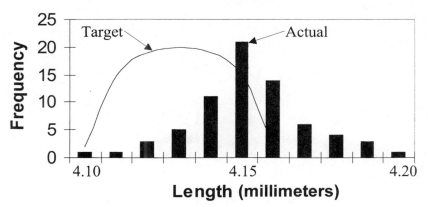

Conformance Graph

change over time. The lines on a control chart provide reference points for use in deciding which type of behavior any given time series displays. When the values stay within the prescribed limits and the data does not indicate that the process may be drifting out of limits, the process or operation being monitored is said to be "in control."

A control chart is simply a time series with three horizontal lines added. A central line is added as a visual reference for detecting shifts or trends, and control limits (computed from the data) are placed equidistant on either side of the central line. The key to the effectiveness of the control chart is the way these limits are computed from the data.

For example, a control chart for the daily percentage of defective parts is shown as Tool 27. The time series there consists of a sequence of single values, starting with data from September 1. As can be seen, this process is "in control," since all of the data points lie between the upper control limit of 20.2 and the lower control limit of 10.2. In other situations, the control chart may be based on a time series of average values or a time series of ranges. While there are several different types of control charts, they are all interpreted in the same way.

When a time series displays unpredictable behavior, the underlying process that gives rise to the time series is said to be "out of control." The essence of statistical control is predictability. A process that does not display a reasonable degree of statistical control is unpredictable.

This distinction between predictability and unpredictability is important because prediction is important to conducting an efficient business. A traditional approach to manufacturing is to depend on production to make the product and on quality control to inspect the final product and screen out items not meeting specifications. It is much more effective to

Control Chart

Daily Percentage of Defective Parts

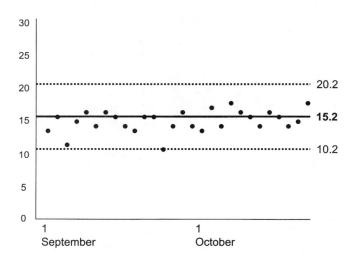

avoid waste by not producing unusable output in the first place—a strategy of prevention. Predictability is a great asset for any process because it makes the manager's job that much easier. When the process is unpredictable, then the time series will be unpredictable, and this unpredictability will repeatedly undermine all best efforts. In fact, attempting to make plans using a time series that is unpredictable, or out of control, results in more frustration than success.

A control chart helps to interpret data. Some of the benefits that can be expected from using control charts are:

- The chart is used to characterize the behavior of the data—is it predictable or not?

- The control chart allows the manager to predict what to expect in the future.

Statistical Process Control (SPC) works effectively to monitor how well a process is able to produce quality results. With SPC, statistically reliable samples of output are periodically measured and charted on a control chart to determine if the process output is within specification. SPC can give early warning that deviations are occurring in the processes.

Questions to consider when a process goes out of control are:

- Has there been a significant change in the environment?

- Has there been a change in the way things are measured?

- Are there differences in methods being used?
- Are untrained workers involved in the process?
- Has there been a change in process inputs?

Use a control chart to:

- Monitor manufacturing processes.
- Identify improvement opportunities.
- Evaluate effectiveness of solutions.
- Monitor administrative and service processes.

Tool 28. Cost- and Revenue-Control Charts

Originally control charts were used almost exclusively in manufacturing. They are now used in many different fields. The cost-control chart is an antidote to the foot-high computerized expense and revenue reports that arrive on manager's desks monthly or weekly in most companies. Depending on the flexibility of your management information systems (MIS) department, you can either ask them to replace the reports with control charts or have your staff summarize the reports in charts for you. Sales managers out in the field can also use cost-control charts to track some of the sales expenses, such as entertainment and travel, that rarely make it into official company reports.

Most companies collect detailed information on marketing expenses, breaking down expenses into sales force, advertising, research, promotion, administrative, and other costs depending on the company and industry. Companies with large sales forces typically generate monthly reports detailing these expenses by territory or area as well. Sales managers often collect statistics on costs incurred by individual salespeople, such as entertainment and travel costs. If these latter costs are not currently available in accounting reports, then they can easily be collected from expense sheets or vouchers.

Here are the four instructions to follow when using cost-control charts.

1. Create a target cost-to-sales ratio for each cost you want to track and for each geographic or functional area you want to track. Translate costs for each period into a cost-to-sales revenue ratio by dividing costs by revenues for the company or, if you are tracking area or territory costs, for the area or territory in question. Second, average a series of cost-to-sales ratios for each cost and for each area or territory. These averages can be used as

targets, or they may be modified slightly to reflect goals based on industry norms if you feel your company's costs are not in line with industry standards.

2. Define a range around the target cost ratio that you consider acceptable. Set both a lower and an upper limit. You can do this statistically—for example, by doing a variance analysis and setting the limits of the range so as to include 95 or 99 percent of the historical figures. You also can set it judgmentally by picking limits that seem reasonable to management. The latter is preferable in my opinion, since the object is to identify ratios that need management's attention—an issue that should be left to management judgment anyway.

   Example: Entertainment expenses typically hover at 5 percent of sales. In some territories, they average one to two percentage points higher. This does not concern management. But every now and then a sales representative goes to town on the company credit cards, pushing a territory's ratio three or more points higher than usual. This does concern management. So management can set the upper limit at 7.75 percent in order to detect these problem situations. For the lower limit, management is concerned that some reps do not entertain customers enough, and wants a tighter range. Therefore, the lower limit will be set at 4 percent. The sales force should be notified of any changes to the targets or the ranges.

3. Create a control chart for each cost ratio and each geographic or functional area you want to track. This control chart can be maintained by staff or can be turned into a computer report. Either way, the system should be designed to provide rapid exception reporting. Whenever a cost ratio moves beyond the upper or lower limit, then management should be notified.

   The cost-control chart (see Tool 28A) consists of a graph with time on the horizontal axis (divided into weeks, months, or quarters depending on management needs and preferences) and the cost ratio on the vertical axis. The target for the ratio is indicated by a bold horizontal line, and the upper and lower limits are also indicated by lines. The chart is updated every period by plotting the latest cost ratio. Whenever the ratio moves beyond the limit, management is notified.

   Note: You also might want to be notified when a cost ratio moves in the *same direction* on the chart for a number of periods in

Tool 28A

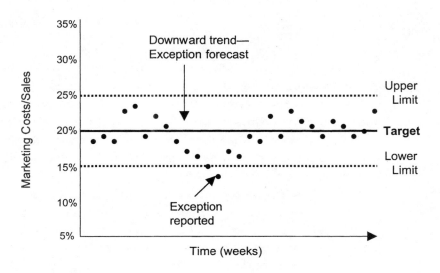

Cost-Control Chart

Tool 28B

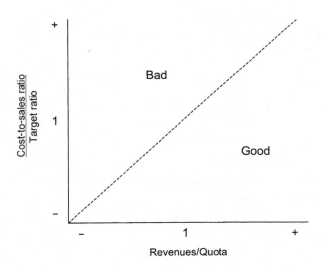

Revenue-Control Chart

a row. The more periods an upward trend is observed, for example, the less likely it is due to chance; it indicates a likely exception in the future.

4. Investigate fluctuations. A control chart is of little use unless reported fluctuations in costs are investigated. Investigation

always should be based on the understanding that costs can fluctuate beyond the control limits due to:

- A real problem or a lack of control over costs
- Chance

Any limit will be exceeded occasionally just by chance, so management must be careful to avoid a punitive tone to investigations. Look for possible causes or underlying problems. Talk to the relevant managers or salespeople. If no good reason for the exception can be found, then see what happens next period. If the exception is repeated next period, then the chances of it being a random occurrence are almost nil.

Idea: Some companies create revenue-control charts as well as cost-control charts, using sales quotas as the target and setting limits a few percentage points above and below. Then sales by territory, area, or even customer can be plotted every period on the revenue-control chart. If your company is using control charts to monitor sales, then consider using them in conjunction with some of the other tools introduced in this book. In particular, create a seasonal index and deseasonalize your sales data (see Tool 20), or create a decomposite graph to separate out the systematic and irregular components of sales (see Tool 21). For example, if you sell plastic rulers you might expect to sell more during the school year and less during the summer months. A control chart would show these normal, cyclical variations in sales as an out-of-control process, because the data points would lie outside of the control limits. If your data is first adjusted to smooth out these normal cyclical variations, then the seasonally adjusted data would describe a process that is by and large in control. Any adjusted data points that lie outside of the control limits are candidates for further investigation.

You can graph both cost and revenue performance of each territory by expressing the cost-to-sales ratio as a percentage of the target ratio and expressing period revenues as a percentage of the target revenues. Then create a graph with expense attainment (as a percent) on the vertical axis and sales quota attainment (as a percent) on the horizontal. (One hundred percent is in the middle of each axis, representing the targets.) A diagonal line from the origin will indicate points at which cost and revenue variations are equivalent. Positions below the diagonal are desirable; positions above it are not. (See Tool 28B.)

Use a cost-control chart to:

- Track sales and marketing costs by product, territory, or salesperson.

- Identify unusually high or low costs for further investigation.

- Create standards for cost-to-sales ratio analysis and computerize the identification and analysis of deviations from the norm.

- Reduce the number of computer forms and statistics that sales managers have to review in order to keep track of expenses.

As a recap, here are the four steps to follow when using a cost-control chart.

1. Create a target cost-to-sales ratio for each cost and geographic or functional area you want to track.

2. Define a range within which cost ratios fluctuate under normal circumstances.

3. Create control charts to graph the fluctuations in cost ratios.

4. Investigate cost ratios that fluctuate beyond the established control limits.

Tool 29. Population Pyramid

The strength of the population pyramid (see Tool 29) is that it is a two-way histogram (included as Tool 13) in which age is plotted on the vertical axis and the number of males and females of each age on the horizontal axis. Females are plotted on one side of the vertical axis and males on the other. Because of space considerations, ages are often grouped into intervals. The most common intervals are two, five, and 10 years. The same quantity scale is used for males and females on the horizontal axis, except they progress in different directions from the central vertical axis. Quantities may be plotted in actual units or percents of the total for a given sex. For example, the percent of females in any given age interval can be calculated by dividing the number of females in that age bracket by the total number of females. Population pyramids provide a graphical means for noting changes or differences in population patterns, either over time within a given population or between different populations.

Population pyramids are useful in business applications to tabulate survey results, or they might be used to help you understand your customer base so that you can target advertising dollars accordingly.

Tool 29

Population Pyramid

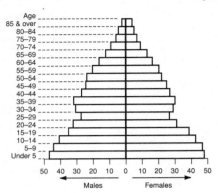

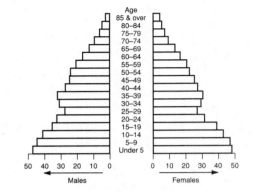

Population pyramid with age scale located on left side

Population pyramid with age scale located in the center

5. Relationships

This chapter contains 12 charts and graphics, including supply and demand, Venn diagrams, and other ways to show relationships.

Tool 30. Supply and Demand Graph

At first glance, the concept of supply and demand appears to be very simple: The more there is of something, the less it costs. This type of graph is used extensively in the field of economics, and virtually every economics textbook contains this graph to convey the fundamental principle of supply and demand. (See Tool 30A.)

Many times a simple graph such as this one can serve as a jumping-off point for more involved decision making. If we were to view the creation of this graph as the beginning of our assignment rather than the end, then what would this graph tell us?

Just as there are two sides to every argument, there are two sides to every market: a *supply* side and a *demand* side. The two sides of the market interact to determine the *price* of the product. This graph tells us the relationship among *supply, demand,* and *price.*

A *market supply line* represents the supply side of a market. It shows the amount of a product that sellers will supply at various prices. For example, the supply line in Tool 30A reveals that the higher the price, the more companies there are willing to supply the product. The position of the supply line depends on a host of factors, including technology and the prices of the resources (labor, capital, and land) used to produce the product.

The demand side of a market can be represented by a *market demand line,* which shows the amount of the product that buyers demand at various prices. The demand line in Tool 30A reveals that the higher the price, the fewer people there are willing to buy the product. The position of this line is dependent on many factors, including consumer tastes, the income level of consumers, the number of consumers in the market, and the level of prices for substitute products.

The *equilibrium price* is the price that can be maintained. The equilibrium price is the price where supply equals demand. Any price that is not at equilibrium cannot be maintained for long, since there are basic factors at work to stimulate a change in price toward equilibrium.

While a supply and demand graph may appear to be very simplistic, its implications for business decision making cannot be underestimated. For example, after using the supply and demand graph as a tool, a manufacturer might make the strategic decision to limit the production of a popular item. Consider how the Ty Corporation's decision to produce limited amounts of certain Beanie Babies pushed prices through the roof, or how limiting the production of the "Charmander" Pokémon card drove up its price.

Tool 30B applies the supply and demand graph to a car manufacturer's analysis of how many of the latest-model-year cars to produce. The overall quantity of cars is determined by how many shifts of production the manufacturer runs at the manufacturing plant. In order to maintain the target sticker price for the car (line S), the manufacturer would run one shift if demand was low (line D low), two shifts if demand was medium (line D medium), and three shifts if demand was high (line D high).

Prices in a capitalistic system are important determinants of what is manufactured, how it is made, and who receives it. Manufacturers cannot produce in an isolated bubble without regard to demand, for supply and demand are inextricably tied.

Tool 31. Yield Curve

A yield curve displays the relationship of interest rates on assets to their time to maturity. Length of time to maturity is shown along the horizontal axis and percent yield along the vertical axis. The vertical scale is always linear with units of percent. The lower and upper values are typically

Tool 30A

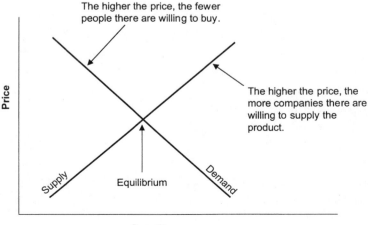

Supply and Demand Graph

Tool 30B

Supply and Demand Graph—Automobiles

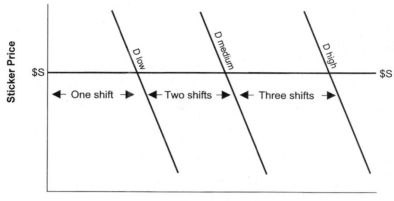

Quantity of Cars/Model Year

slightly below and above the lowest and highest values plotted, respectively. The placement of values on the horizontal axis varies widely. Sometimes values are placed uniformly. Other times there is no apparent pattern. As a result, curves of the same data can look quite different.

When the percent yield is higher on assets with longer-term maturities, then the curve is called a positive yield curve. (See Tool 31A.) When the percent yield is lower on assets with longer-term maturities, then the curve is referred to as a negative or inverted yield curve. (See Tool 31B.)

Companies may use yield curves in their treasury departments, depicting anything from bonds (long or short term) to certificates of deposit, Treasury bills, bank accounts, or money markets, to see how efficiently they are using their money.

Tool 32. Break-Even Graph

A break-even graph is used to estimate when the total sales of a company equal the total costs of the company: the break-even point. The same concept and graph can be applied to a product, a type of service, or any facet of a business where sales and fixed and variable costs can be identified and the variable costs calculated on a per-unit basis. One of the main purposes in establishing the break-even point is to determine the volume of business a company must do to begin making a profit. In theory, the graph is valuable and easy to use and understand. In practice, it is not easy to apply with a high degree of precision due to the difficulty of accurately estimating and allocating each of the values.

Tool 31A

Positive Yield Curve

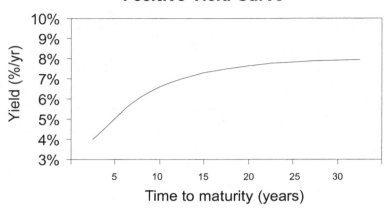

Long-term interest rates higher than short-term rates

Tool 31B

Negative (Inverted) Yield Curve

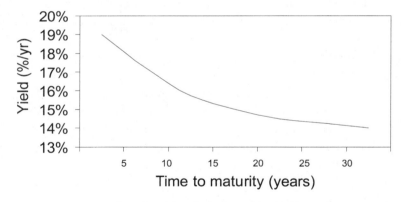

Long-term interest rates lower than short-term rates

Units (e.g., pieces, volume, time) are generally shown on the horizontal axis. In Tool 32, total sales dollars are shown on the vertical axis. Total sales typically include such things as product mix, various discount schedules, returns, and the like. The other variable plotted is total costs, both direct and indirect. The categories of fixed and variable costs may differ from application to application. The total sales and total costs lines intersect at the break-even point, beyond which the company or product being studied theoretically begins making a profit. In a theoretical break-even graph, the lines generally are drawn straight. In practice, these lines might be straight, curved, or stepped.

Break-Even Graph

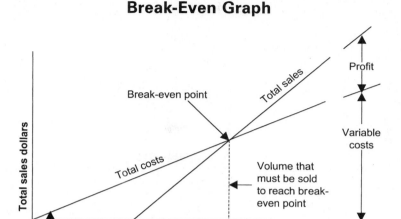

Tool 33. Experience Curve

An experience curve is sometimes referred to as a learning curve. An experience curve shows how the cost of producing products decreases as cumulative output increases. Why does cost decrease with accumulated experience? Because as a company makes more of a product, it becomes smarter about how to make it. This intelligence leads to increased productivity and efficiency, which results in cost reductions.

An experience curve demonstrates that the unit cost of value added to a standard product declines by a constant percentage (typically between 20 percent and 30 percent) each time cumulative output doubles. The importance of experience curves is in focusing on the rate and sources of experience-based cost reductions.

Tool 33 shows a classic industry experience curve of the cost per ton of steel decreasing as a function of the tons produced. Experience curves can be found for individual companies over time and for entire markets. This example includes two curves, one showing the overall U.S. market, and the other specific to a single steel producer (SBJ Technology).

Tool 33

Experience Curve

Steel Costs, 1990–1996

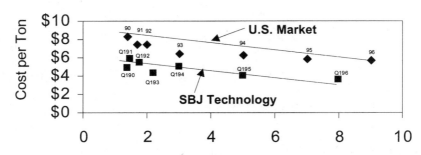

Accumulated Experience (Sum of Tons Produced)

Tool 34. Venn Diagram

Venn diagrams are tools used in many different fields including math, psychology, education, advertising, and sociology. They are used for such diverse purposes as studying complex concepts, introducing young children to math, analyzing interrelationships between groups, and illustrating ideas in presentations.

Venn diagrams typically are used to describe a relationship between two or more sets of things or information. They accomplish this by the relative positioning of circles, each representing a set of information. For example, two circles might be used to depict two sets of buyers. One circle

Tool 34A

Venn Diagram

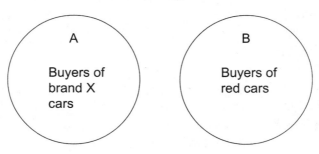

Circles used to represent one set of buyers that purchased brand X cars and another set of buyers that bought red cars

Tool 34B

Venn Diagram—Overlap

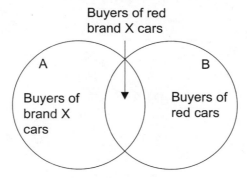

Tool 34C

Venn Diagram—Two Levels of Overlap

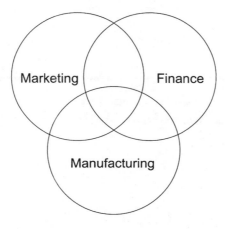

Tool 34D

Venn Diagram—No Overlap

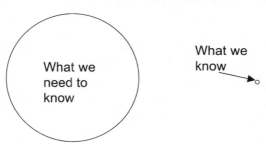

Tool 34E

Specialty Division Data Flow Chart

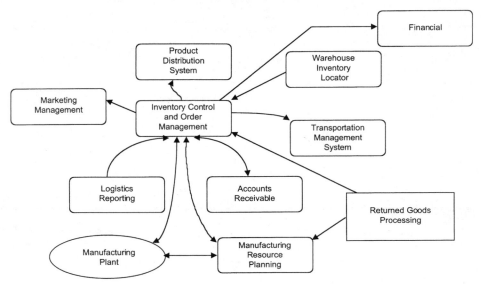

Tool 34F

Venn Diagram—
Management Information Systems

indicates the set of buyers that bought brand X cars from a given dealer, and the other circle, the set of buyers that bought red cars from the same dealer. (See Tool 34A.) To convey the idea that some of the same buyers appear in both circles (i.e., they bought red brand X cars), the two circles are drawn overlapped. (See Tool 34B.)

Tool 34C is an example of a Venn diagram with three symbols and two levels of overlapping. One level of overlap indicates that each pair of departments shares responsibility in some areas. The other level indicates that all three departments share responsibility in other areas.

Tool 34D depicts two divergent sets. These circles do not overlap; or, said another way, the intersection of these two sets is the null set. A Venn diagram can depict items that do not overlap.

Another example of a practical use of Venn diagrams for presentation purposes is in the area of management information systems (MIS). MIS is one of the most challenging areas finance professionals are facing. A variety of factors, including changing technologies, requests for information, or rapid growth, may indicate the need to reengineer these systems. It can be difficult to explain to senior management that the company's systems need to be reengineered.

When it comes to making a presentation on systems, less may be more. Generally it is more productive to provide only the most essential information. A detailed data flow chart may be a required first step for the systems analyst reengineering the flow of data within an organization. A detailed data flow chart, however, can be complex. (See Tool 34E.) While some senior executives may be interested in reviewing such a chart, more likely most will want simply a few key "takeaways."

The presenter may want to make the case that one critical component of the total MIS needs to be replaced. Although this component interfaces with several other computer systems, there is no need to describe all of the system links.

The Venn diagram simply and effectively communicates the overall systems environment. (See Tool 34F.) The sizes of the circles vary in proportion to the importance of each individual component, and the overlaps of the circles convey the extent of the relationships among them. Most senior executives would prefer a concise and crisp analysis of a situation to achieve a better understanding of it overall. This is a classic case of how selectively showing less data can provide more information.

Tool 35. Fishbone Diagram (Ishikawa Diagram)

A fishbone diagram is sometimes referred to as an Ishikawa diagram, a cause-and-effect diagram, or a characteristic diagram. A fishbone diagram is a method for systematically reviewing all factors that might affect a given objective or problem. It is a graphic tool used to explore and display opinion about the sources of variation in a process. The method was developed by Kaoru Ishikawa, the father of Japanese quality control, as an aid in analyzing a process and identifying the factors that need to be controlled to improve the quality of the process.

The purpose of a fishbone diagram is to arrive at a few key sources that contribute most significantly to the problem being examined. These sources are then targeted for improvement. The diagram also illustrates the relationships among the wide variety of possible contributors to the effect. This is accomplished by assigning each factor to a line or arrow and arranging the lines and arrows into meaningful clusters in a hierarchical fashion. The lines and clusters are arranged so that they all feed into a central line that, in turn, feeds into a symbol representing the overall object or problem. (See Tools 35A and 35B.)

Start creating a fishbone diagram by entering the name of a basic problem of interest at the right of the diagram at the end of the main "bone." Then draw the main possible causes of the problem as bones off of the main backbone. The "Four M" categories can be used as a starting point: "Materials," "Machines," "Manpower," and "Methods." Different names can be chosen to suit the problem at hand, or these general categories can be revised. The key is to have three to six main categories that encompass all possible influences.

The fishbone diagram can be used by individuals or teams, but probably it is used most effectively by a group—engineers, production workers, or others. Sometimes the group that does the analysis is a quality control (QC) circle, but it can also be used by an ad hoc committee if your manufacturing facilities do not use QC circles.

Typically, a team leader first presents the main problem (far right of main bone) and draws the diagram on a whiteboard. The leader then asks for assistance from the group to determine the main causes, which are subsequently drawn on the board as the main bones of the diagram. Brainstorming is typically done to add possible causes to the main "bones." This subdivision into ever-increasing specificity continues as long as the problem areas can be further subdivided. The practical maximum depth of this diagram is typically four or five levels. Because it encourages brainstorming, a fishbone diagram is a highly creative tool.

The team assists by making suggestions, and, eventually, the entire fishbone diagram is filled out. When the fishbone is complete, a detailed picture of all the possible root causes of the designated problem exists. At this point, team discussion takes place to decide the most likely root

Tool 35A

Fishbone Diagram

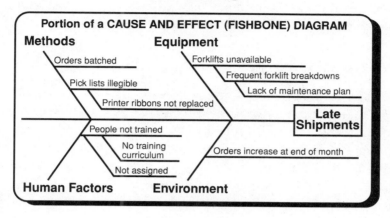

Portion of a CAUSE AND EFFECT (FISHBONE) DIAGRAM

Tool 35B

Ishikawa Diagram

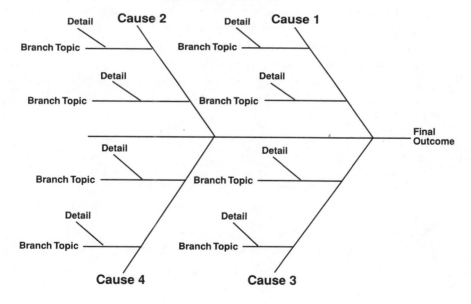

causes of the problem. These causes are circled to indicate action items, and the use of the tool is complete.

The fishbone diagram is a visualization and knowledge organization tool. Simply collecting the ideas from a group in a systematic way facilitates understanding and ultimately diagnosing the problem. Use the fishbone to sort out and identify possible root causes of a problem. Bear in mind that while a fishbone diagram and the accompanying cause-and-effect analysis can help zero in on likely root causes, they do not necessarily confirm them. You may need to gather more data to verify that you have uncovered and classified the true root causes.

Tool 36. Affinity Diagram

An affinity diagram is a group decision-making technique designed to sort a large number of ideas, process variables, concepts, and opinions into naturally related groups. These groups are connected by a simple concept. (See Tool 36.) An affinity sort is the first step in developing an affinity diagram. It is a technique to categorize ideas into subgroups.

Here are the guidelines to follow to make an affinity diagram.

Conduct a brainstorming session on the topic under investigation. Clarify the list of ideas. Ensure ideas are described with phrases or sentences. Record them on small cards or Post-it notes. Aim for five to 10 groups of ideas. If one group is much larger than the others, consider splitting it. Minimize discussion while sorting—discuss while developing header cards (discussed below). Randomly lay out cards on a table, flipchart, or wall. Without speaking, sort the cards into "similar" groups based on your gut reaction. If you do not like the placement of a particular card, move it. Continue until consensus is reached. Create header cards consisting of a concise three- to five-word description—the unifying concept for the group. Place a header card at the top of each group. Discuss the groupings and try to understand how the groups relate to each other.

Tips:

- Ask whether ideas are adequately clarified.
- Use only three to five words in the phrase on the header card to describe the group.
- If possible, have groupings reviewed by nonteam personnel.
- While sorting, physically get up and gather around the area where the cards are placed.
- Team members will ultimately reach agreement on placement, if for no other reason than exhaustion.
- Sorting should not start until all team members are ready.
- If an idea fits in more than one category or group, and consensus about placement cannot be reached, then make a second card and place it in both groups.

Tool 36

Affinity Diagram

| | | MAINTAINING A SUCCESSFUL QUALITY PROCESS | | |

Tool 37. Mileage Table

Anyone who has ever used a road map knows that a mileage table clearly shows in table format how many miles there are between major cities. A mileage table differs from the other charts and graphs in this book because it has the same headings for both the rows and the columns.

For example, in Tool 37A, we see the cities of Cedar Rapids, Davenport, Des Moines, Dubuque, and Sioux City in both the top row and the left-hand column. To find the distance between Cedar Rapids and Des Moines, we can find Cedar Rapids in the top row and Des Moines in the left-hand column. Then we see where the two intersect by traveling down the Cedar Rapids column until we meet the Des Moines row. We can see that there are 122 miles between these two cities. Similarly, if we search for Cedar Rapids in the left-hand column and Des Moines in the top row, we can also see that there are 122 miles between the cities. Finally, notice that wherever a city's row intersects that same city's column, there is a blank. These blanks represent 0 miles; they bisect the table along the diagonal, and they divide the table into two triangles.

The knowledge flow grid (see Tool 37B) is an application of the mileage table to business. A knowledge flow grid is a compact, matrix representation of how knowledge flows in an organization. This example contains all of the core processes of a health maintenance organization (HMO), which are depicted in order in the top row from left to right and in the left-hand column from top to bottom. These seven processes are:

1. Business Planning

2. New Product & Service Introduction

3. Build Delivery System & Back Office

4. Acquire Customers & Members

5. Installation & Data Maintenance

6. Member Health Care Enhancement

7. Patient Health Care Delivery

Much like a mileage table, the grid is intended to show how one place relates to another: that is, how information flows from one process to another. The knowledge flow grid provides insights about how to manage a complex organization such as an HMO and highlights issues of information needs and requirements.

In this HMO example, where there are seven core processes, the number of interfaces requiring evaluation is 42. (A 7 x 7 matrix has 49 squares, but, following the same logic as a mileage table, seven of them are blank.) In the mileage table, we are indifferent as to whether we look at the mileage from Cedar Rapids to Des Moines or from Des Moines to Cedar Rapids. In either case, the mileage is the same. In the knowledge grid, however, we look at two distinct connections.

1. The "feed forward" of information downstream

2. The "feed back" of information upstream

The blocks below the diagonal in the lower triangle represent the *forward* flow of information; there are 21 possible intersections for all the "feed-forward" relationships. An example of a feed-forward relationship is when a lower-number process relates to a higher-number process: for instance, the relationship between Business Planning and Acquire Customers & Members. The blocks above the diagonal in the upper triangle reveal a *feed back* from a later (i.e., downstream) process to an earlier (i.e., upstream) one. This means that the earlier process has to be revisited in light of the late arrival of new information. An example of a feed-back relationship is when a higher-number process relates to a lower-number process: for instance, the relationship between Patient Health Care Delivery and Build Delivery System & Back Office. There are 21 possible intersections for all the feed-back relationships.

At the core process level, each intersection deserves a look. At first it sounds unwieldy to list each core process and determine how knowledge might flow to and from each. However, this grid provides the means to perform such an evaluation. While the intent should be to evaluate every single box on the grid, only three examples are displayed in the HMO grid.

1. **Business Planning → Acquire Customers & Members.** It is the planning process that defines and selects specific customer segments as targets. Once defined, and as these targets are continuously refined, the information needs to be fed forward to the acquisition process. The acquisition process then can build in specific points where selected customer segments are whisked through the process, while the nontargeted customer segments are encouraged to go elsewhere.

2. **Installation & Data Maintenance → Member Health Care Enhancement.** The intent of the Member Health Care Enhancement process is to prevent a member from becoming a patient. Therefore, it is imperative that member-specific health information, which enables preventive health care, be fed forward to drive specific kinds of preventive care. For instance, a member with a family history of heart disease might require more frequent and specialized physical examinations.

3. **Patient Health Care Delivery → Build Delivery System & Back Office.** As claims are processed, it is not unusual to have a significant portion "suspend" for one reason or another. These suspended claims then require labor-intensive intervention to process. To reduce the delays and extra labor, information regarding the portion of suspended claims that is due to inherent system deficiencies needs to be fed back to the systems development

Tool 37A

Mileage Table

	Cedar Rapids	Davenport	Des Moines	Dubuque	Sioux City
Cedar Rapids		79	122	70	275
Davenport	79		168	70	349
Des Moines	122	168		200	200
Dubuque	70	70	200		402
Sioux City	275	349	200	402	

Knowledge Flow Grid

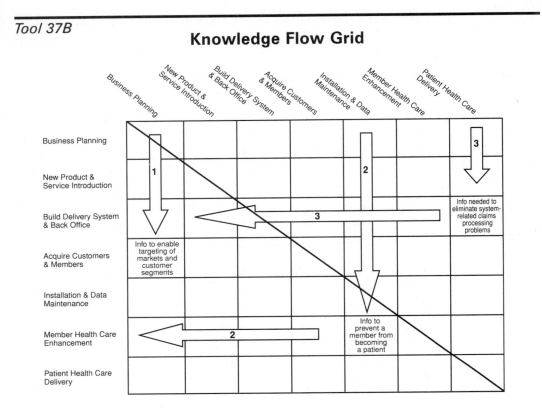

operation. The information is vital to determining priorities relative to other problems and to designing and implementing the fix. The key is in defining where the information is most vital and where it truly becomes valuable at a specific point of use in the process.

The intent of the knowledge flow grid is to force an evaluation of each one-to-one link among the core processes. This evaluation means first defining the key information needs and content for every intersection on the grid.

Tool 38. Journeys-Between Chart

The journeys-between chart is a simple, manual tool for planning the optimal layout of production equipment and functions. It helps the manager or engineer minimize movement of loads between nonadjacent departments and reduce the overall time wasted in material movement. Alternative approaches involving complex computerized optimization programs are often not worth the time and effort and are harder to adapt to a process's unique (and changing) requirements and constraints. This method is a practical tool for the manager who needs a usable evaluation of the current setup in a hurry. (See Tool 38.)

For example, imagine that you work in a manufacturing plant where you receive truckloads of raw materials through a large receiving dock. In order to produce your finished goods, 90 percent of the raw materials that you receive need to be first weighed and then sorted. Obviously, it would make the most sense to have your incoming inspection area (e.g., scales and sorting machines) located near the receiving dock. Optimally, your raw materials would flow from station to station around the shop floor in the most efficient manner possible. Often the work flow is determined by habits, even if these habits are out of date and inefficient. The journeys-between chart can be part of an overall effort by you and your employees to eliminate wasted time and to improve productivity.

Idea: The journeys-between chart is also useful for improving the efficiency of data processing, clerical, order fulfillment, parts warehousing, and many other processes in a company where transfers of material take significant time and energy. For example, the chart might be adaptable to an airline to understand its work flow better by tracking the movement of an individual ticket. Upper management can encourage a department to adopt the more rigorous approaches of manufacturing by requesting a chart from the process manager and defining operational efficiency goals by using a chart-based statistic, such as percent decrease in nonadjacent material movements.

Here are the three steps to follow to create a journeys-between chart.

1. List each separate department or process. Create a table with the full list across the top (with the label "To") and again down the side (with the label "From").

2. Fill in the chart by counting the number of separate loads, no matter what they are, that move between any two points. Enter this total in the appropriate cell of the chart. Do this count for a fixed period of time—10 minutes, an hour, or a day—whatever provides a good sample of work-flow patterns for the process under investigation. You may need several observers or several observation periods if the process is complex or the facility is large.

3. Enter the totals onto a map or diagram of the facility to see whether the highest numbers of movements are between adjacent points or between distant points. For a more rigorous analysis, determine the distance or average travel time between each point and multiply this by the number of journeys. Sum the weighted

Tool 38

Journeys-Between Chart

Station:

From \ To	A	B	C	D
A		25	10	6
B	25		8	100
C	10	8		35
D	6	100	35	

Station:

Number of Directional Journeys Between Each Station Pair

measures for an estimate of total time required for material movement. This statistic can be used to set efficiency goals, to evaluate alternative process configurations, and even to create competition between different plants producing the same product.

Use a journeys-between chart to:

• Measure the efficiency of a facility or process.

• Redesign a manufacturing process to reduce the time wasted on moving materials between departments.

• Make quick design changes in response to the changing requirements of a batch or to-order facility.

• Apply principles of efficient manufacturing to administrative and operational processes in an organization.

Tool 39. Alignment Graph

An alignment graph is designed to solve an equation involving two parameters and one variable. Such graphs consist of three or more scales arranged so that a straight line would intersect all the scales at values that satisfy the equation.

The alignment graph shown as Tool 39 is a paper version of a slide rule. The left column (in a log scale ranging from 1 GB to 1 kB) lists various sizes of information packets that one might want to send electronically. On the rightmost column (also in log scale) are the times it might take for transmission, ranging from a second to a week. In between these two columns is a third column, representing various modes (speeds) of transmission. To use this graph, place a ruler or straightedge to connect three dots, one from each column. This alignment graph allows one to

Tool 39

Alignment Graph

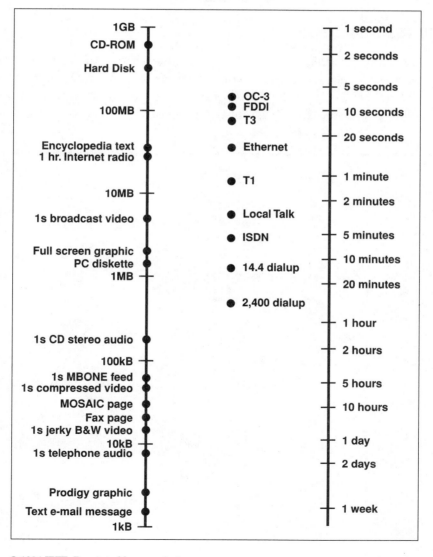

© 1994 IEEE. Reprinted by permission.

choose the size of the information package to be sent from the first column, pair it with a mode of transfer from the second, and read off the shipping time from the third. A text e-mail message over a 2,400-baud line takes less than 10 seconds, a fax page at 2,400 baud just under two minutes. Moreover, one can choose various permutations by linearly joining any two and reading off the third. For example, we can use the graph to decide on carriers. If we are backing up hard disks often over a network, then it tells us we should not consider anything slower than Ethernet.

This graph is quite satisfactory being left implicit, since the general story is obvious. Big things take longer to send; faster carriers send things more quickly.

Tool 40. Belt Chart

A belt chart looks like a bull's-eye, with each ring of the bull's-eye representing a data series. Since a belt chart contains more than one data series, it enables the viewer to look concurrently at the distribution of related data in a number of different ways. By doing this, patterns and relation-

Tool 40

Belt Chart

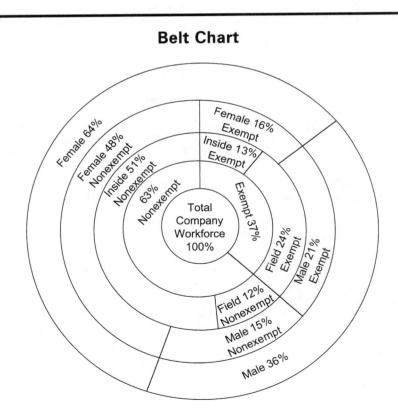

ships become clearer. (See Tool 40.) The following are representative of the types of observations that can be made by this particular belt chart. In the workforce:

64 percent are female, 36 percent are male

63 percent are nonexempt, 37 percent are exempt

48 percent are nonexempt females, 15 percent are nonexempt males

21 percent are exempt males, 16 percent are exempt females

64 percent work in the office, 36 percent work in the field

Tool 41. Trilinear Graph

Trilinear graphs are used to plot information that has three variables. The sum of the three variables can be any value; however, it is generally one or 100 percent. A typical example is the percent of material, labor, and overhead in the total cost of a product. Different products have different percentages of the three elements, but the percentages of the three elements for most products add up to 100 percent.

A trilinear graph consists of an equilateral triangle (all three sides of equal length) with each line from a vertex to the opposite base representing one of the three variables. Scales are distributed along the altitudes

Tool 41A

Trilinear Graph—General

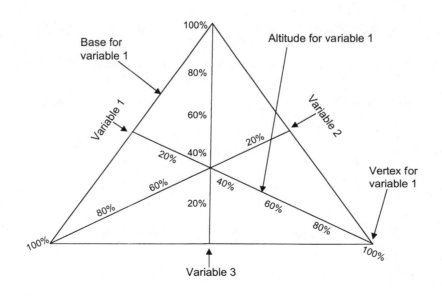

Trilinear Graph—Specific

Material, Labor, and Overhead Content for a Product

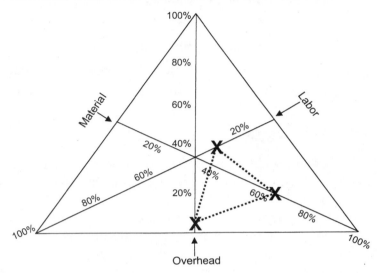

with zero at the base and 100 percent at the vertex. Tool 41A depicts in general terms the format of a trilinear graph.

Tool 41B is a specific trilinear graph that depicts the material, labor, and overhead content for a product. The labels at each base tell you which component (i.e., material, labor, or overhead) is being measured. The altitude tells you the percentage for that component. In this example, the product is comprised of 65 percent material cost, 30 percent labor cost, and 5 percent overhead cost. In order to understand where its money is being spent, a manufacturing company might want to make specific trilinear graphs for each product that it manufactures and then compare the graphs.

6. Presentation

Ten techniques that have a wide variety of applications are presented here.

Tool 42. Pie Chart

A pie chart consists of a circle divided into wedge-shaped segments. It shows the proportional size of items that make up a data series to the sum of the items. Its major purpose is to show the relative sizes of components to one another and to the whole. Pie charts are used extensively as communication tools in presentations and publications.

The portion of the circle included between adjacent radii is called a segment, slice, sector, or wedge. It may be measured in terms of angle, area, or arc. The size of the segment is in the same proportion as the data element it represents.

Our example, Tool 42A, shows the distribution of different types of vehicles sold during 2000. Since 50 percent of the vehicles sold are cars, the segment of the pie representing cars constitutes 50 percent of the total pie. The 50 percent can be determined in any of three different ways, all of which yield the same end result. They are:

1. The angle of the wedge (i.e., 50 percent of 360 degrees)

2. The area of the wedge (i.e., 50 percent of the total area of the circle)

3. The length of the arc (i.e., 50 percent of the circumference of the circle)

The same reasoning applies to trucks and vans. Twenty-five percent of the vehicles are trucks and 25 percent are vans; therefore, 25 percent of the area of the pie is allocated each to trucks and vans. The sum of all three segments (which adds up to 100 percent) makes up the complete pie representing all of the vehicles. In our example, all of the segments are labeled with a description of what the segment represents, the actual value of the segment, and the percent the segment represents of the whole.

To provide additional detail for a slice, you can break down that item in a smaller pie or bar chart next to the main chart. See Tool 42B, in which the total number of vans sold is broken down into the number of new vans sold (35) and the number of used vans sold (15).

Tool 42A

Pie Chart

Motor Vehicles Sold by Dealer X in 2000

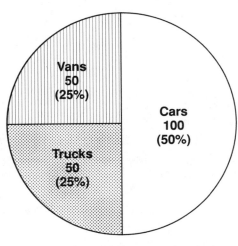

Total vehicles sold equals 200

Tool 42B

Pie Chart—with Detail

Vehicles Sold in 2000   ### Vans Sold in 2000

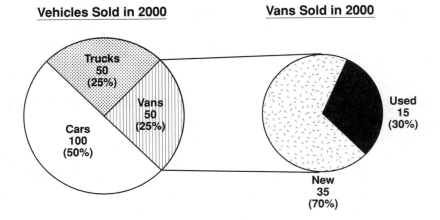

Total vans sold equals 50

Some general characteristics of pie charts are:

- The major advantage is that people naturally tend to think that a circle represents 100 percent. Percentages over 100 percent can be used but rarely are, since they go contrary to one of the major advantages of pie charts.

- From a technical point of view, the segments can be arranged in any order. To make the chart meaningful and easy to read, the segments are typically arranged in some meaningful order, such as smallest to largest, natural groupings of the data, or alphabetically.

- Information can proceed in a clockwise or counterclockwise direction. The clockwise direction is normally used.

- The starting or reference radius can be located at any point around the circle. It frequently is located at 12 o'clock.

- Color is particularly important in pie charts.

- With rare exceptions, negative numbers are not displayed on pie charts.

Tool 43. Doughnut Chart

You might think that the doughnut chart is used by hungry police officers, but it is not. A doughnut chart is a pie chart with an area blanked out in the center so information, such as the overall value of all the pieces of the pie, can be shown. One of the criticisms sometimes expressed about pie charts is that they focus on the relative sizes of the components to one another and to the whole but give no indication of changes in the whole when two or more pie charts are shown. The doughnut chart partially addresses this issue by somewhat more forcefully bringing changes in overall values to the viewer's attention.

As seen in Tool 43, the chart reader can now see not only how each segment has changed, but also what that means relative to the total amount.

Tool 44. Area Graph

An area graph can be considered as much a process or a technique as a graph type, since area graphs are generated by filling the area between lines generated on other types of graphs. For example, when only one data series is involved, filling the area between the data curve and the horizontal axis converts a simple line graph (see Tool 44A) into an area graph (see Tool 44B). Area graphs generally are not used to convey specific values. Instead, most frequently they are used to show trends and

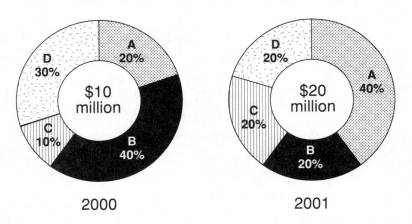

Tool 43

Doughnut Charts

Sales by Product (A, B, C, and D)

2000   2001

relationships, since they emphasize the magnitude of change over time. Tool 44B represents a typical application of an area graph in which average electric power usage is recorded for a 24-hour period.

A circular area graph is an area graph wrapped into a circle. (See Tool 44C.) Sequential, time series scales frequently are used on the circular axis. The quantitative value being measured is shown on the radius or value axis. Tools 44A, 44B, and 44C all depict the same data (i.e., the average electric power usage for a 24-hour period).

Tool 45. Rose Chart

A rose chart is a circular graph with its circumference divided into equal-size segments (i.e., segments with equal angles). The number of segments is equal to the number of data elements in the data series. Each segment of the circle has a sector plotted in it, the relative size of which is proportional to the data element it represents. Categories are distributed equally around the circumference of the circle, and values are plotted along the radii, with zero at the center of the circle. An example of a rose chart is shown as Tool 45.

Rose charts differ significantly from pie charts (included as Tool 42). The wedges of a pie chart generally all have the same radii, and the sizes of the segments in a pie chart are varied by changing the angle of each. In a rose chart, all sectors have the same angle, and the relative sizes of the sectors are varied by changing the radii. Rose charts can be constructed

Tool 44A

Simple Line Graph

Average Electric Power Usage

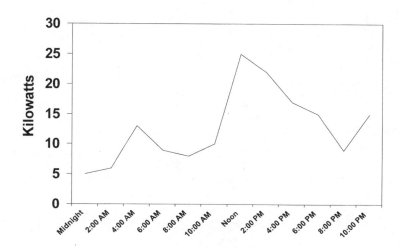

Tool 44B

Area Graph

Average Electric Power Usage

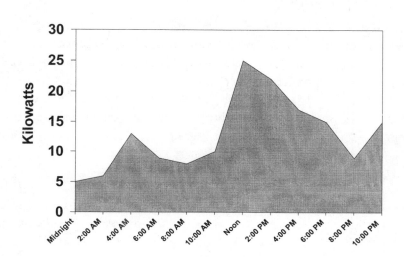

Tool 44C

Circular Area Graph

Average Electric Power Usage

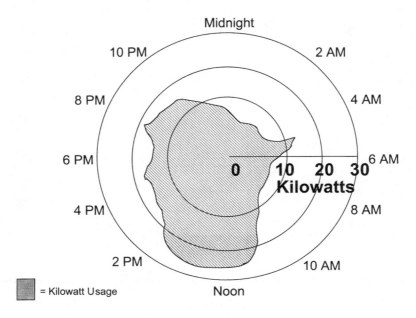

= Kilowatt Usage

so that either the areas or the radii of the sectors are proportional to the values they represent.

Tool 45 shows monthly sales data; for example, sales in the month of June were $20 million. While the information contained on a rose chart can be displayed using other chart types, sometimes it helps to view information (particularly cyclical information) in this format. Varying your repertoire is the graphical equivalent of thinking outside of the box. If you include imaginative graphics in a presentation, your viewers might remember more of your message.

Tool 46. Vector Chart

A vector chart uses the angle and/or size of arrows to communicate qualitative information. The vector chart in Tool 46 relates products and markets. In it the angle of the arrow describes the nature of a market. An arrow straight up indicates that market is doing very well and is growing rapidly. Straight down means the market is doing poorly, pointing slightly down means it is shrinking slightly, and level means the market is static.

Vector charts often are constructed in the form of a matrix or table.

Tool 45

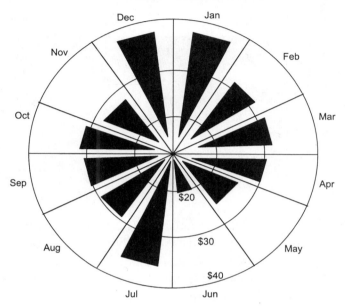

Rose Chart

Monthly Sales ($ Millions)

The items being reviewed are distributed in rows and columns in some meaningful fashion. The only guidelines regarding placement of headings in rows and columns is that similar things be handled uniformly. For instance, if time periods are one of the variables, and it is decided to make them into column headings, then all time periods should be displayed as column headings and distributed in chronological order. Specific information is encoded into the matrix by placing a vector (arrow) in each of the cells.

This chart would be useful to a product manager when deciding how to allocate resources among product lines. While direction (angle) of the arrow is important, so too are the total size and growth rate of the market(s) for the product.

Tool 47. Scatter Diagram

When ALCOA needed to know if it was using its money wisely, a scatter diagram helped answer the question.

In the late 1980s, senior managers at the worldwide, integrated aluminum company became increasingly concerned that it might not be getting the maximum benefit from its fixed asset investments because of

Tool 46

Vector Chart

Market Sector

		New Housing	Home Repair	Rental Property	Office Property
Product	A	↗	↑	↘	↗
	B	↘	↗	↘	↘
	C	↗	↘	↗	↘
	D	↗	↘	→	↘

less-than-optimal capital expenditure decisions. (ALCOA, in 1989, had capital expenditures of about $900 million.) In addition, the process of allocating capital to the businesses within ALCOA was perceived to be flawed. Various groups within the company presented a series of studies that reinforced these concerns. One of these studies involved a scatter diagram.

Tool 47 shows ALCOA and other selected companies on a scatter diagram. The vertical axis measures a 10-year average return on equity (ROE), while the horizontal axis depicts 10 years of capital expenditures as a percent of the same 10 years of sales. The companies in the bottom-right quadrant have lower returns, and they spend a significantly higher portion of their resources on new equipment and facilities (or, alternatively, generate less revenue from their equipment and facilities). Obviously, it is better for a company to be in the top half of this diagram with a higher ROE. The benchmarked companies demonstrate that it is difficult to be in the top half of this diagram if capital expenditures as a percent of sales are greater than 9 percent. This finding was confirmed by analyzing the largest 800 industrial companies in the Compustat database, a Standard & Poor's library of financial and market information of 6,000 American companies. Only rarely are companies in the top-right quadrant.

Its placement in the bottom-right quadrant of the scatter diagram concerned ALCOA. To reach the ROE targets ALCOA had set for itself, managers had to either reduce the long-term level of capital expenditures with the same revenue or greatly increase the revenue generated by the existing level of capital expenditures.

This scatter diagram was a very powerful tool for ALCOA. In general terms, scatter diagrams are a family of graphs that display quantitative information by means of dots. These graphs generally have quantitative scales on both axes and can accommodate many data points. They are used extensively for exploring relationships, particularly correlations between two or more data sets, for example, the relationship between efficiency and speed, or dollars spent on food vs. dollars of income, or (as we have seen in our example) 10-year average ROE vs. 10 years of capital expenditures as a percent of the same 10 years of sales. After data is plotted on a scatter diagram, patterns formed by the data points are used to make observations about the relationships of the data sets graphed. In ALCOA's case, these observations led to the creation of a high-level team to study the capital expenditures decision process. This team improved the way ALCOA analyzed and justified capital projects. The bottom line for ALCOA is that it now makes its fixed asset investments more wisely.

Tool 47

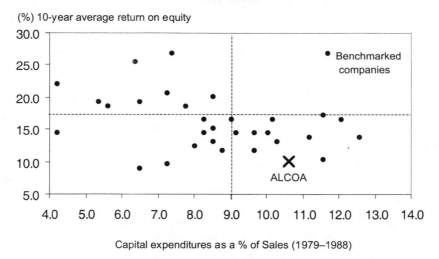

Tool 48. Bubble Graph

A bubble graph is a variation of a scatter graph where the data points (dots) have been replaced by circles (bubbles of various sizes). The bubble graph compares sets of three values: one value is displayed by the bubble's location on the X axis, one by its location on the Y axis, and the third by the size of the bubble—proportional to the value it represents.

Obviously, bubble graphs have an advantage over point or line graphs because they can display one additional variable in the same space. Tool 48A is a bubble graph where the X axis represents the breadth of a company's product line, the Y axis is a measurement of quality, and the diameters of the circles are proportional to sales.

There are several ways to explain the bubbles' quantitative information: A legend can be provided, values can be shown in or near the bubbles, or a sample-size bubble can be included for reference (as we have done in Tool 48B). In this example, each bubble's location represents a company's quality information and number of products sold. The size of the bubble indicates the value of the third variable: the company's market share measured in percent. The chart in this example shows that while Company G has the most products and the greatest market share, it does not have the highest quality.

How to re-create the bubble graph (market share study—Tool 48B) using Excel

Enter the data shown in columns A through D, and then click the Chart Wizard button. Select Bubble as the chart type, and then click Next. Set the data range to cells B2:D8, with the series set to columns. Then click Next to go to step 3 in the Chart Wizard dialog, and you will be prompted to set the chart options. Label the chart title, X axis, and Y axis as shown. Under the Legend tab, remove the check mark for Show Legend.

Click Next again to go to step 4, and select the chart location to be as an object in your current spreadsheet. Click Finish to view the chart. Then, using the mouse, you can adjust the size and position the bubble graph to the appropriate location.

To complete the formatting, double-click on the Y axis, and in the Scale tab of the Format Axis dialog, set Minimum to 0, Maximum to 6, and Major Unit to 1. Set number format to number and decimal places to 0. Double-click on the X axis, and in the Scale tab, set Minimum to 5. Apply other formatting as desired. For example, you can display the labels as shown to identify which bubble represents which company. Go to Chart Options and activate Show labels under the Data Labels tab. Then change the labels by clicking on each data label and typing in the company name. We have also added a sample-size bubble in the upper-right corner by clicking on Insert, Picture, AutoShapes, selecting an oval, and then drawing a circle of the same diameter as the circle for Company G with 35 percent market share. We then labeled the sample-size bubble for reference.

Tool 48A

Bubble Graph

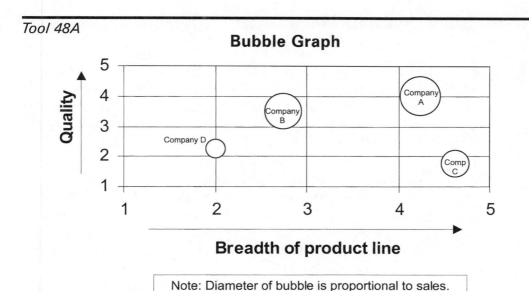

Breadth of product line

Note: Diameter of bubble is proportional to sales.

Tool 48B

Bubble Graph—Market Share Study

	A	B	C	D	E	F	G	H	I
1		Number of products	Quality	Market share %					
2	Company A	14	2	14					
3	Company B	20	5	24					
4	Company C	18	2	5					
5	Company D	8	1	6					
6	Company E	17	1	5					
7	Company F	19	4	11					
8	Company G	24	3	35					
9				100%					

Industry market share study

Tool 49. Stoplight Chart/Moon Chart

Although some of the greatest creative minds in the world's history have studied and used astrology, a moon chart has nothing to do with that subject. The next examples are called stoplight charts or moon charts because they use symbols that resemble stoplights or the phases of the moon to get their point across.

Tool 49A contains some of the more commonly used stoplight chart and moon chart symbols. Now let us see these symbols in action.

Tool 49B (stoplight chart) depicts frequency-of-repair records for automobiles built during the late 1990s. It contains multiple repetitions of these small symbols. This is a particularly ingenious mix of table and graphic, portraying a complex set of comparisons between manufacturers, types of cars, year, and trouble spots. It is chock-full of information yet easy on the eyes and easy to digest.

Tool 49C (moon chart) conveys a situation analysis and competitive landscape. This tool is wonderful for competitive analysis and/or acquisition analysis.

Tool 49D (moon symbol on a comparison chart) is a feature comparison chart for electric drills. This type of display can be a powerful marketing tool, and it also can be helpful in decision making.

Tool 49A

Stoplight Chart and Moon Chart Symbols

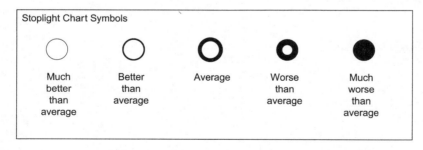

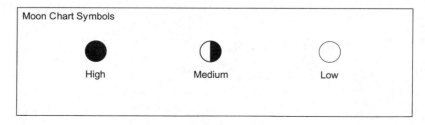

Tool 49B

Stoplight Chart

Frequency-of-Repair Records for Automobiles

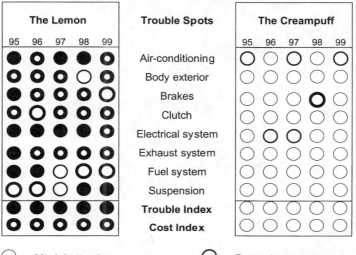

	The Lemon					Trouble Spots	The Creampuff				
	95	96	97	98	99		95	96	97	98	99
	●	◉	●	●	◉	Air-conditioning	○	○	○	○	○
	◉	◉	◉	⭕	◉	Body exterior	○	○	○	○	○
	●	◉	◉	◉	⭕	Brakes	○	○	○	**⭕**	○
	◉	⭕	◉	◉	◉	Clutch	○	○	○	○	○
	●	●	●	●	◉	Electrical system	○	**⭕**	**⭕**	○	○
	●	◉	◉	◉	◉	Exhaust system	○	○	○	○	○
	●	●	⭕	⭕	⭕	Fuel system	○	○	○	○	○
	⭕	⭕	⭕	●	●	Suspension	○	○	○	○	○
	●	●	●	●	●	**Trouble Index**	○	○	○	○	○
	◉	◉	◉	◉	◉	**Cost Index**	○	○	○	○	○

○ = Much better than average ◯ = Better than average

⭕ = Average ◉ = Worse than average ● = Much worse than average

---

Tool 49C

Moon Chart

Situation Analysis and Competitive Landscape

	Company A	Company B	Company C	Company D
Scale and Size	●	◗	◗	○
Industry Experience	●	○	◗	●
Structure Complexity	○	●	○	○
Leverage	●	◗	◗	◗

Key:

● High   ◗ Medium   ○ Low

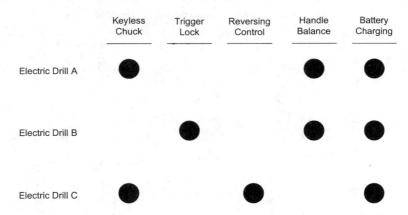

Moon Symbol on a Comparison Chart

	Keyless Chuck	Trigger Lock	Reversing Control	Handle Balance	Battery Charging
Electric Drill A	●			●	●
Electric Drill B		●		●	●
Electric Drill C	●		●		●

Tool 50. Chernoff Faces

Chernoff faces are simplified, cartoonlike faces that can be used to graphically display complex multivariate data. They draw on the human mind's innate ability to recognize small differences in facial characteristics and to assimilate many facial characteristics at once. The method was developed by the Stanford statistician Herman Chernoff and presented in a 1973 paper in the *Journal of the American Statistical Association* titled "The Use of Faces to Represent Points in k-Dimensional Space Graphically." If you want to know more about Chernoff faces, I encourage you to read this article. Although academic, it is quite readable by those without a background in statistics.

Chernoff faces provide a graphical technique for encoding multivariate information (generally three or more variables) into the facial features of small icons so that the viewer gets an overview of the data based on facial expressions. For example, the position of the eyes might be proportional to one variable (e.g., looking to the left indicates a large number, to the right a small number, and straight ahead an average number). The size of the ears might be proportional to another variable (e.g., big ears, 100,000; little ears, 1,000). Both quantitative and qualitative information can be encoded. Chernoff faces may be used as stand-alone images, in matrixes, or in conjunction with another type of chart, such as a map or graph. In this example (see Tool 50A), the financial performance of four companies is compared by showing four faces, one for each company. Each company's financial data has been encoded into the features of the face representing that company. The four faces are then used as

Tool 50A

Chernoff Faces—Graph

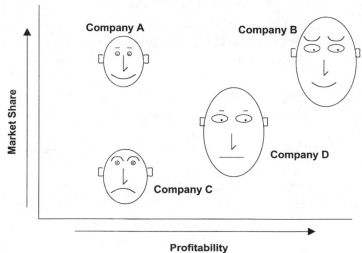

Tool 50B

Chernoff Faces—Map

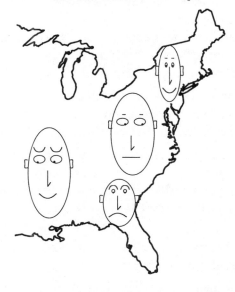

symbols on a graph. The next example (see Tool 50B) is a map that shows another variation in which faces represent data for different regions.

If, during the encoding process, care is taken to ensure that positive values are associated with the more pleasant facial features, then the viewer can get a general feel for each entity based on whether the face is happy or sad.

Tool 51. Logarithmic and Semilog Graph

A logarithmic graph has one or more logarithmic scales. A logarithm is the power to which a base number must be raised in order to equal a given value. For example, the logarithm of 100 using a base of 10 is two since 10 must be raised to the power of two (10^2) to equal 100.

If a graph has one logarithmic scale and one nonlogarithmic scale, it is called a semilogarithmic or semilog graph. On semilog graphs, the logarithmic scale is generally on the vertical axis. If the graph has two logarithmic scales, then it is called a full logarithmic graph.

Use full logarithmic and semilog graphs to emphasize nonlinear relationships, such as economies of scale or percentage rate changes. When data forms a straight line on a semilog grid, then the data along the logarithmic axis is increasing at a constant percentage rate while the variable on the other axis is increasing linearly. A straight line on a full logarithmic grid means the data along both axes is increasing at constant percentage rates.

The steepness of the curve on a semilog graph at any point is proportional to the actual rate of change of whatever is being plotted. The steeper the slope, the greater the rate of change. A downward slope indicates a negative rate of change.

Tool 51 is a business application of a semilog graph. It contains four curves: sales in dollars, sales in units, rejects, and profit. Here is how to interpret the curves on this graph.

- When the curve is a straight line (sales in dollars), then the change is taking place at a constant rate, such as 20 percent per year for each successive year.

- When the curves are parallel (sales in dollars and sales in units), then the rates of change are the same for the two data series, even though the actual values are different. In this example, the rates of growth for the product are the same (20 percent per year) both in terms of units and dollars.

- When the slope of the curve becomes shallower as it progresses (rejects), then the rates of change are decreasing.

- When the slope of the curve becomes steeper as it progresses (profit), then the rates of change are increasing.

Daily stock prices are often plotted on semilog graphs. If the stock price tends to fluctuate around a straight line with an upward slope, then the price of the stock, on average, is increasing at a constant rate comparable to the percent represented by the straight line.

Tool 51

Semilog Graph

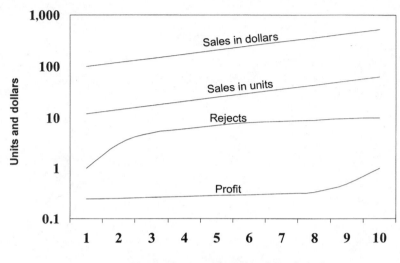

7. Value Ranges

This chapter covers confidence intervals, high-low graphs, and candle-stick charts.

Tool 52. Confidence Interval Chart

A confidence interval is sometimes referred to as possible margin of error. In terms of charts and graphs, a confidence interval graphically indicates the range of values within which a given value probably exists. For example, if an organization wants to estimate the percent of members that will participate in a given activity, the organization might send a questionnaire to 10 percent of the members asking if they plan to participate. If 15.2 percent of the people receiving the questionnaire indicate they will participate, it may be assumed that about 15.2 percent of all of the organization's members will participate. Even though it is highly unlikely the overall participation will be exactly 15.2 percent, one can calculate with a given degree of confidence, such as 90 percent, the range of values within which that overall participation will probably lie. This range of values is called a confidence band or confidence interval. If the range of values extended from minus 3.4 percent to plus 3.4 percent of the sample value, then the confidence interval extends from 11.8 percent (15.2 percent – 3.4 percent) to 18.6 percent (15.2 percent + 3.4 percent).

A confidence interval can be shown graphically several different ways. (See Tool 52.) The top and bottom of confidence intervals are generally called the upper and lower confidence limits, respectively. When confidence intervals are shown, two additional bits of information are sometimes included.

1. The level of confidence associated with the interval: For example, if the confidence level is 50 percent, then there are 50 out of 100 chances that the actual value will lie within the confidence interval. Frequently used confidence percentages are 50, 68, 90, 95, and 99.

2. The reasons for the potential spread in values: These may be due to sample size, sample-to-sample variations, biases of the people being interviewed, or methods of processing the data.

Tool 52

Confidence Interval Chart

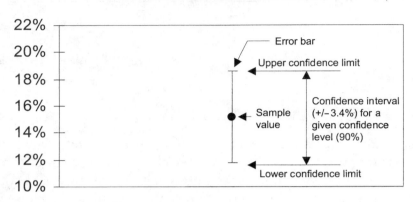

Tool 53. High-Low Graph

The high-low graph is used frequently for recording the price of stocks. Each symbol represents a set of data, with the top representing the highest value, and the bottom, the lowest value. (See Tool 53.)

This graph might be used to compare the range of your company's stock price to that of your competitors. Doing this might be helpful when explaining to shareholders why your stock experienced large fluctuations in price over a given period of time. Perhaps certain macroeconomic factors affected not only your company, but also most, if not all, of your competition.

For example, most airlines operating in the same geographic area would feel the impact of a significant rise in fuel costs coupled with an unusually snowy winter season. In this scenario, a high-low graph showing the stock prices of five competitors would reveal dramatic highs and lows before even accounting for an individual company's performance.

Tool 54. Candlestick Chart

Candlestick charts are used to record and analyze the selling prices of stocks, bonds, and commodities, with particular attention to price reversals. (See Tool 54.) A chart might represent the stock price of a single entity (General Electric, IBM, McDonald's) or the average of multiple entities (all stocks in a particular exchange, selected industrial companies, selected utilities). Price is measured along the vertical axis, and uniform intervals of time progress along the horizontal axis from left to right. Time intervals range from minutes to years. Days, weeks, and months are

Tool 53

High-Low Graph

Stock Prices

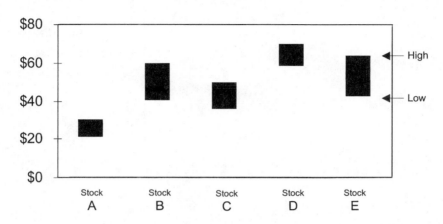

among the most frequently used. Each symbol represents a given period of time. Vertical price scales can be shown on the left side, the right side, or both sides.

The basic candlestick symbol consistently displays opening price, closing price, highest price, and lowest price for the applicable period. The symbol consists of a rectangle with a straight line extending from the top and bottom. If the closing price is lower than the opening price, then the rectangle is *black*. If the closing price is higher than the opening price, then the rectangle is *white*.

If the rectangle is black, then the *top* of the rectangle designates the *opening* value and the *bottom* of the rectangle designates the *closing* value. The black body of the symbol is a strong graphical indicator that the closing price of the stock during that period was lower than the opening price.

If the rectangle is white, then the *top* of the rectangle designates the *closing* value and the *bottom* of the rectangle designates the *opening* value. The white body of the symbol is a clear graphical indicator that the closing price of the stock during that period was higher than the opening price.

The *top* of the vertical line designates the *highest* price for which that particular stock sold in that period of time. When there is no line, then the high is designated by the top of the rectangle. The *bottom* of the vertical line designates the *lowest* price for which that particular stock sold in that period of time. When there is no line, then the low is designated by the bottom of the rectangle.

Tool 54

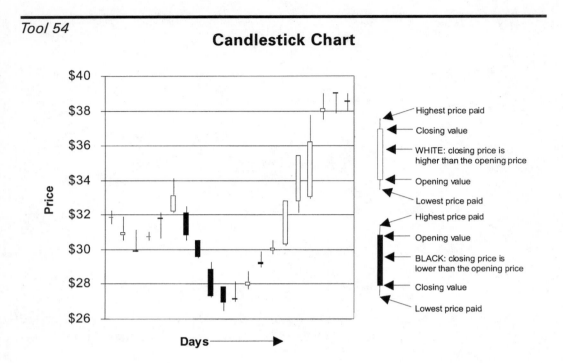

Candlestick Chart

8. Schematics and Maps

This chapter covers schematics, process diagrams, and maps.

Tool 55. Block Diagram

A block diagram uses geometric figures and symbols (referred to as blocks) to lay out schematically systems, networks, concepts, circuits, procedures, and structures. Block diagrams also are used for planning, developing, communicating, and organizing thoughts. A block diagram for starting a company might have five blocks representing five key elements. Those elements might be developing the idea or concept; raising capital; assembling a management team; acquiring facilities, product, and a workforce; and initiating the business. If these five elements are enclosed in rectangles (see Tool 55A), the resulting graphic might be called a simplified block diagram (even though the blocks are not connected). When these blocks are used for planning purposes, a group of people might shuffle them around until they agree on the best sequence for a particular situation. Once agreement is reached, the blocks might be arranged in their proper sequence and connected by arrows. (See Tool 55B.)

When used in this way, block diagrams function as an aid in the planning, development, and communicating processes by providing a graphical means of addressing major elements of a project without too much detail. The blocks on a diagram may have different shapes to indicate type of function, facility, organization, and the like, and may be connected with one or more lines or arrows. (See Tool 55C.) In Tool 55C, a rectangle may represent a step in a process, while the triangle and the circle may represent decision points and the end of the process, respectively. When the purpose of the block diagram is simply to indicate such things as relative physical location or the existence of marginally related ideas, or information, there may be no interconnecting lines. (See Tool 55D.) Block diagrams generally address the broad overview of a subject. When they become detailed descriptions of the subject, many times they are referred to as flow charts, process charts, or procedural charts.

Tool 56. Flow Chart

A flow chart is a pictorial representation that displays interrelated information such as events or the steps in a process. A flow chart can be constructed to depict the following.

- The sequence of steps in a process and the relationships between them
- The procedure for getting a request approved for capital expenditures
- How paperwork progresses through an organization
- Currency moving from country to country and bank to bank
- Products being manufactured
- The development of an idea

The things being represented can be tangible or intangible. Flow charts can be very general, for overview purposes, or very detailed, with lots of supporting and auxiliary information for use in day-to-day activities. They can describe an entire sequence of events from start to finish or address just a portion of the overall sequence.

When used by a team to review a process, the steps in the process are presented graphically in an organized fashion so that team members can examine the order presented and come to a common understanding of how the process operates. Flow charts can be used to describe an existing process or to present a proposed change in the flow of a process.

Flow charts are the easiest way to "picture" a process, especially if it is very complex. Flow charts should include every activity in the process. A flow chart should be the first step in identifying problems and targeting areas for improvement.

Tool 55A

Block Diagram—Simplified

Develop idea or concept	Assemble management team

Raise capital	Acquire facilities, product, and workforce	Initiate business

Block diagram indicating the five major
elements in starting a business

Tool 55B

Block Diagram—with Arrows

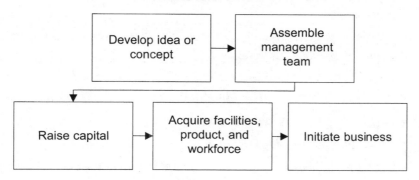

Block diagram indicating the five major elements in starting a business

Tool 55C

Block Diagram—with Different Shapes

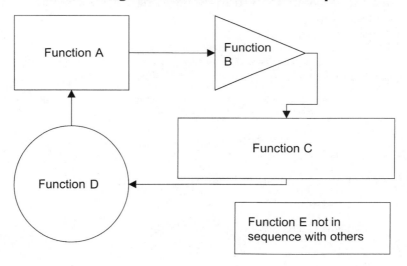

Tool 55D

Block Diagram—Unconnected

| Engine compartment | Passenger compartment | Luggage compartment |

Tips:

- Determine the appropriate starting and ending boundaries of the process.

- Do not flow-chart too big a process. Break it down into subprocesses.

- Ask: "What happens next?", "Is there a decision to be made at this point?", "Does this reflect reality?", "Who else knows this process?"

- Keep the flow chart clear and simple.

- Verify the process with those who perform the activities.

- Show all decision steps, feedback loops, and wait loops.

- Look for ways to simplify, shorten, and improve the process being flow-charted.

When possible, do a walk-through of the process to see if any steps have been left out or extras added that should not be there. The key is not to draw a flow chart representing how the process is supposed to operate but to determine how it actually *does* operate. A good flow chart of a bad process will show how illogical or wasteful some of the steps or branches are.

Use flow charts to:

- Document, analyze, or develop a path of activities or steps in a process, either actual or ideal.

- Describe processes, ideas, or networks, particularly complex and abstract ones.

- Uncover delays and non–value-added activities as part of process improvement efforts.

- Improve communications.

- Aid in troubleshooting.

- Help to clarify ideas.

- Serve as a tool in planning and forecasting.

- Reduce misunderstandings and conserve time.

- Simplify training.

- Document procedures.

- Illustrate cross-functional relationships and responsibilities.

The most widely used types of flow-charting symbols are shown in Tool 56A. Tool 56B is a flow chart of process improvement. Tool 56C is a flow chart of how ideas get to market.

Tool 56A

Flow-Charting Symbols

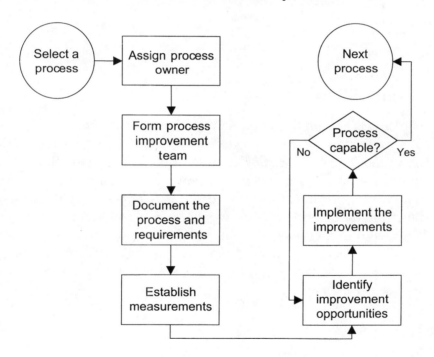

| Start and end | Steps in | Decisions in | Flow of |
| of process | process | process | process |

Tool 56B

Flow Chart of Process Improvement

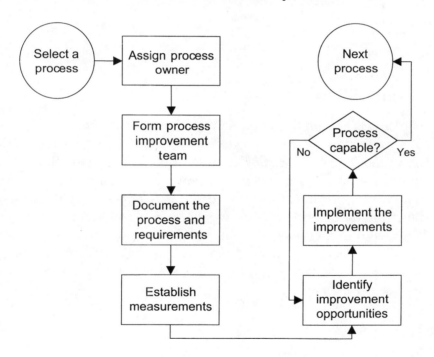

Tool 56C

"Idea to Market" Flow Chart

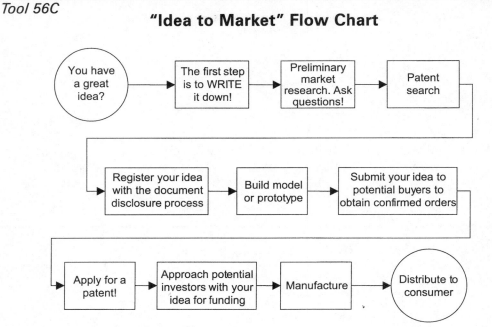

Tool 57. Route Map

Sometimes seemingly simple solutions provide the clearest picture of all. The route map for the airline (see Tool 57) clearly shows the destination of all the flights originating from the hub. Route maps show the paths things take to get from one place to another. The paths or routes might be walkways, corridors, roads, railroad tracks, streets, shipping lanes, overhead cables, or buried pipes. Extensive amounts of information can be encoded into the lines used for route maps, such as type of pavement, number of lanes, diameter of pipe, voltage of cable, and the like. When route maps connect multiple locations, they sometimes are called networks or web maps.

Tool 58. Flow Map

Route maps show the paths from one point to another but generally do not indicate what or how much moves or flows in what direction along the paths. Flow maps, on the other hand, say little or nothing about the exact path but do indicate such things as:

- What is it that flows, moves, or migrates?
- What direction is the flow moving or what are the source and destination?

Route Map

- How much is flowing, being transferred, or transported?
- Does the map provide general information about what is flowing and how it is flowing?

Flow maps can be used to show movement of almost anything, including tangible things such as people, products, produce, natural resources, and weather, as well as intangible things such as know-how, talent, credit, or goodwill. Sometimes the subject of a flow map is indicated in the title of the map. In other cases, it is noted directly on the map, as shown in Tool 58. When many different things are flowing, a legend can be helpful.

How precisely the direction and location of flow is indicated may range from very general to very detailed. In some situations, such as a railroad, flow is concentrated with limited options as to the number of directions it can move. In other cases, the movement is distributed over a large area, with movement sometimes occurring in many directions at once. The direction of the movement may be constant or in flux. In these cases, special data graphics sometimes are developed, such as in a weather map. Arrows are used almost universally to indicate the direction of the flow.

Tool 58

Flow Map

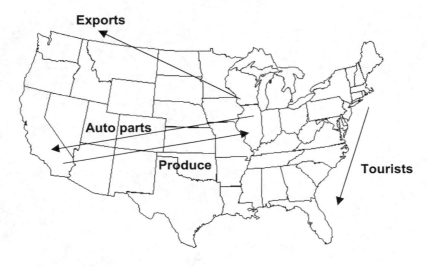

Tool 59. Distorted Map

A distorted map is sometimes referred to as a proportional or value-by-area map. A distorted map is a map in which the sizes of the entities on the map are proportional to values other than the actual surface areas of the entities. For example, if a particular resource is being studied by country, then the size of each country on the map is proportional to the availability of the resource in that country. The country with the largest supply of that resource is drawn with the largest area on the map even if it has the smallest physical area in terms of actual land. (See Tool 59A.)

There are no well-established guidelines as to how to construct distorted maps; therefore, there exists an almost limitless number of possible variations. Tool 59B shows three different examples of New England to give some idea of how diverse designs can be. For example, these graphics might be used to display the number of cows, horses, and chickens in each state.

Another type of distorted map keeps the shape of the area unchanged and simply varies the size of the entities in proportion to the values they represent. Doing this generally requires that there be no shared boundaries between the areas (i.e., a noncontiguous map), as shown on the right-hand side of Tool 59C. Even though the entities are not contiguous, their relative locations are maintained as near as possible. Viewers can recognize the entities more easily if their original shapes and relative locations are retained.

Tool 60. Prism Map

A prism map is a variation of a topographical map in which areas are elevated in proportion to the values they represent. For example, the height of the various states might be proportional to the grain produced in each of the states. The more grain produced per acre, the taller the prism representing that state. Because of the difficulty of accurately estimating actual values, this type of map is generally more qualitative than quantitative.

Tool 59A

Distorted Map—Africa

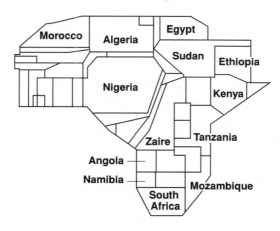

Tool 59B

Distorted Map—New England

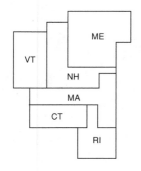
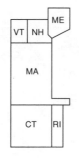
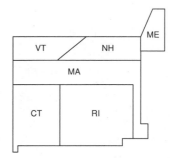

Distorted variations of the same six states.
Different data is used in each map.

Tool 59C

Distorted Map—Before and After

Undistorted for
reference

Sizes proportional to
values they represent

Each area (prism) of a stepped three-dimensional map might have its unique height based on its actual value, or class intervals might be established and discrete heights assigned to each class interval. The actual shape of the area being represented is generally shown; however, the areas can be distorted to encode a second variable. Tools 60A and 60B are examples of undistorted and distorted prism maps. A legend normally is required to aid the viewer in decoding the map. Grid lines, tick marks, and scales are seldom used.

Tool 60A

Undistorted Prism Map

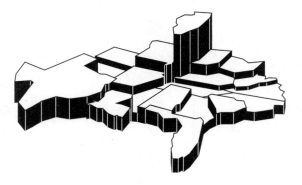

Areas on the map have the same
shapes as the areas they represent.

Tool 60B

Distorted Prism Map

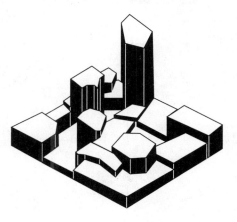

Areas on the map are distorted
and do not have the same shapes
as the areas they represent.

Tool 61. Web Map

The next example appeared in *Fortune* magazine and shows the links among the 10 most admired companies. At a glance, it shows which companies have the most connections with others. (See Tool 61.)

This chart was created by *Fortune* magazine's senior graphic designer, Linda Eckstein, who says this about producing clear and striking business graphics: "It's easier to create better graphics when the accompanying article contains great ideas." She achieves clarity in her graphics by eliminating unnecessary detail and by distilling the ideas down to their essence. It is during this process of deciding what to eliminate that, according to Eckstein, her designs virtually reveal themselves.

Tool 61

Web Map

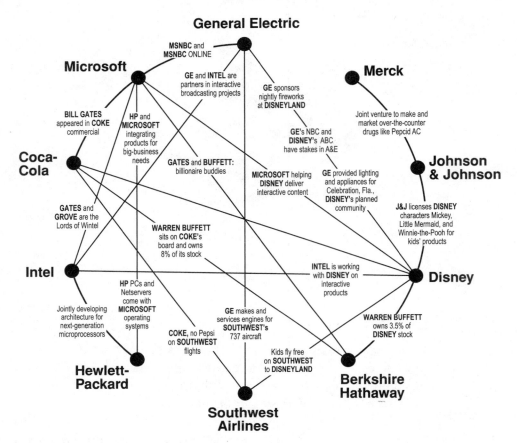

General Electric

Microsoft

MSNBC and
MSNBC ONLINE

GE and INTEL are
partners in interactive
broadcasting projects

GE sponsors
nightly fireworks
at **DISNEYLAND**

Merck

BILL GATES
appeared in **COKE**
commercial

HP and
MICROSOFT
integrating
products for
big-business
needs

Joint venture to make and
market over-the-counter
drugs like Pepcid AC

GE's NBC and
DISNEY's ABC
have stakes in A&E

Coca-Cola

GATES and BUFFETT:
billionaire buddies

MICROSOFT helping
DISNEY deliver
interactive content

GE provided lighting
and appliances for
Celebration, Fla.,
DISNEY's planned
community

Johnson & Johnson

GATES and
GROVE are the
Lords of Wintel

WARREN BUFFETT
sits on **COKE**'s
board and owns
8% of its stock

J&J licenses **DISNEY**
characters Mickey,
Little Mermaid, and
Winnie-the-Pooh for
kids' products

Intel

INTEL is working
with **DISNEY** on
interactive
products

Disney

Jointly developing
architecture for
next-generation
microprocessors

HP PCs and
Netservers
come with
MICROSOFT
operating
systems

GE makes and
services engines for
SOUTHWEST's
737 aircraft

WARREN BUFFETT
owns 3.5% of
DISNEY stock

COKE, no Pepsi
on **SOUTHWEST**
flights

Kids fly free
on **SOUTHWEST**
to **DISNEYLAND**

Hewlett-Packard

Berkshire Hathaway

Southwest Airlines

© 1998 Time Inc. Reprinted by permission.

9. Organizing Data and Information

Structure charts and other ways to organize data are presented here.

Tool 62. Organization Chart

Organization charts are diagrams that show how people, operations, functions, equipment, activities, and the like are organized, arranged, structured, and/or interrelated. They apply to any size of organization. A typical organization chart consists of text enclosed in geometric shapes that are connected with lines or arrows. Charts of this type generally progress from top to bottom.

Tool 62A contains information on major organizational units. Charts with this type of information show the fundamental structure of an organization. They are applicable to any type of organization, whether a corporation, a government, a nonprofit organization, or the military.

Tool 62B contains information by title. This type of chart often is used to convey information about function as well as title and frequently is used in conjunction with the names of individuals holding the titles. When used to compare various organizations, charts with titles sometimes indicate more about personnel philosophies than about the structure of the organization. For example, given the exact same position and responsibilities, one organization might give the title of vice president, while others might give titles such as manager, director, or supervisor.

In some cases, individuals report to two or more people for various aspects of their responsibilities. For example, the financial manager in a division might report directly to the general manager of the division and indirectly to the corporate financial manager. Or a coordinator of new product development might report directly to the engineering manager and indirectly to the manufacturing manager. A widely accepted method to depict this on an organization chart is to show the direct reporting relationship with a solid line and the indirect relationship with a dotted line. (See Tool 62C.)

Organization charts serve many different purposes. These include:

- Defining lines of authority and responsibility
- Showing who reports to whom

- Showing how people, departments, organizations, concepts, or equipment interrelate
- Making it known where to go to resolve problems or concerns
- Orienting people to a system or organization
- Assisting in the analyzing and administering of large organizations
- Providing a tool for use in the planning process
- Providing a tool for comparing the structure of various organizations

An organization chart can be used to explain complex measurements. By using a standard format in a fresh way, an analyst can take a complicated concept such as economic value added as the starting point of an organization chart and break it down into its value drivers. (See Tool 62D.)

When the key topic is strategy, organization charts can be used to condense a large amount of data into a concise, single-page format. (See Tool 62E.) Corporate strategies have many components that balance the competing interests of customers, employees, and shareholders. These components are held together by a single vision or "driving force" that usually is articulated by the CEO and that might include strategies for finance, marketing, manufacturing, and research and development. For these components to be integrated effectively in the strategic planning process, a large volume of ideas must be organized and complex issues analyzed and resolved. In this example, each objective then may be used as a key topic for individual departments to develop and to implement.

The strategic plan should articulate clear long-term objectives from which specific near-term goals are derived. The way to achieve these goals and objectives most effectively is by using an organization chart. Just as an organization chart arranges a company into divisions, it also can sort related ideas into groupings. In addition, the visual display of relationships helps ensure that the corporate vision together with the related objectives and strategies are in alignment. The organization chart format ensures completeness, tests for reasonableness, and highlights interrelationships and dependencies among the elements of the strategic plan.

Most presentation software packages can produce Tool 62E easily. The corporate vision has been placed in the box where the president is listed on a management structure organization chart. Reporting directly to the corporate vision, in the boxes usually reserved for vice presidents, are the objectives. The objectives' "direct reports" are the specific strategies that will sustain the objectives over time, eventually realizing the corporate vision.

Finance professionals often play a key role in the strategic planning process by integrating detailed divisional plans into a master plan. One of the most valuable elements of this master plan is a written document summarizing each divisional or departmental strategy and its interrelationships.

The organization chart with its display of various plan components is not a substitute for the subjective assessment of central issues and marketplace realities. Moreover, the chart is by no means the only effective way to summarize a strategic plan. Rather, it demonstrates how to adapt one readily available tool to a new application. The end product is a one-page summary that clearly identifies vital issues, ensuring that no important elements are overlooked and helping a company achieve its goals.

When the topic at hand is more modest than establishing and reviewing corporate strategy, the organization chart format still comes in handy. Tool 62F is a personal budget organization chart. This chart progresses from left to right.

Tool 63. Strategic Issue Analysis Map

William King, a professor at the University of Pittsburgh's Graduate School of Business, developed the strategic issue analysis map to improve on an organization's ability to make judgments about important issues and to integrate issue analysis into the regular planning process. His method helps managers and analysts cooperate and achieve consensus. It also can be adapted to individual decision making, as the discipline it provides is valuable.

Note: This strategic planning method does not produce any prescriptions. It collects and organizes information for use in decision making but does not have any specific formulas or models of the decision process. For those executives who want to make their own strategic decisions, this method provides a good vehicle for utilizing staff time and expertise in a helpful but nonintrusive manner.

Here are the six instructions to follow to use the map.

1. Management lists strategic issues. These are important situations in which outcomes are uncertain enough that analysis is indicated. Formally, they are defined as:

   - Having outcomes important to the organization's performance
   - Having uncertain or controversial impact
   - Requiring different strategies depending on their outcome

Tool 62A

Organization Chart—Divisional Structure

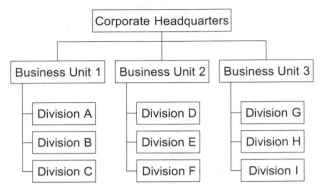

Tool 62B

Organization Chart—Management Structure

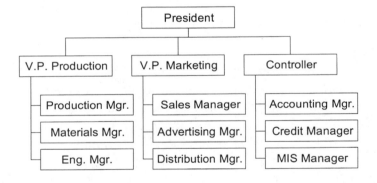

Tool 62C

Organization Chart—Management Structure
(with dotted-line responsibility)

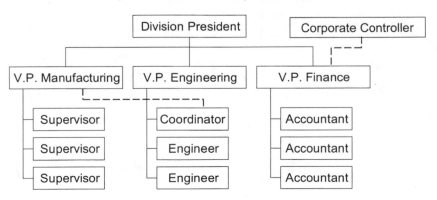

Tool 62D

Organization Chart—Measurements

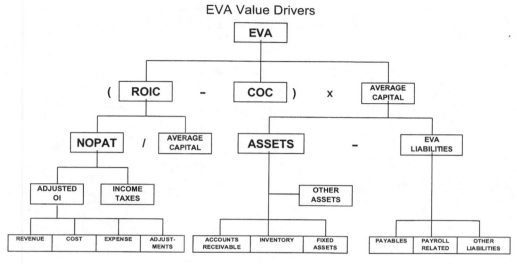

EVA Value Drivers

Key:
EVA = Economic Value Added
ROIC = Return on Invested Capital
COC = Cost of Capital
NOPAT = Net Operating Profit after Tax
OI = Operating Income

Tool 62E

Organization Chart—The Strategic Plan

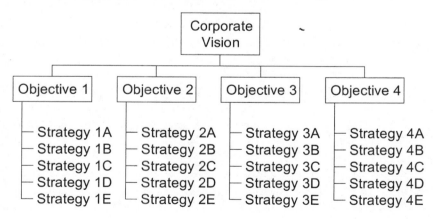

Organization Chart—Personal Budget

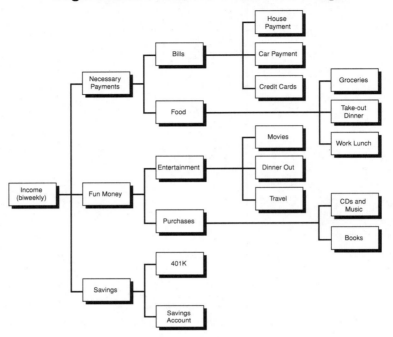

The question "Will there be a recession next year?" would be a strategic issue for most companies, especially if a recent slowdown in the economy made recession a real possibility. The outcome is likely to be important to an organization's performance, but the exact impact is probably uncertain, and different strategies would be required depending on whether there will be a recession and how deep and long it lasts.

2. Management and staff cooperate to make a formal statement of the issue. Clarifying questions are explored. This step ensures that analysts understand management's concerns, and it adds depth and clarity to the issues.

3. Analysts build a strategic issue analysis map—a diagram illustrating all the subissues that relate to the strategic issue. The theory behind it is that any big question can be better understood by breaking it down into a series of small questions.

   Our example is of a firm that is concerned that another may enter its market. The strategic issue in this case is the threat of entry. The extent and nature of the threat can be understood only by looking at such questions as whether there are economies of scale in this market, how much capital it takes to enter, whether the new competitor would have any cost advantages, and so forth. Some of these subissues may have to be broken down even further, as is cost advantages in the example. (See Tool 63.)

4. Management reviews the model to improve or clarify it, making sure all important subissues are identified. This step allows management to examine the analytical model before staff time is invested in analysis and makes sure it is focused on valuable issues.

   Idea: If resources are limited, prioritize the issues and explore the most important. For example, if the firm concerned with the threat of entry thinks its potential competitor has a cost advantage, then it would see this as a key issue needing immediate analysis and verification.

5. The previous steps lead to agreement on specific subissues that require detailed research. Now analysts or staff from appropriate departments are delegated the task of information gathering. Questions that require expertise or time unavailable within the organization can be farmed out to consultants or research firms. Relevant information is added to the strategic issue diagram or put into a briefing paper. This information and the issue map itself form an information model that provides a solid foundation for management judgment and decision making.

6. Now managers can use the strategic issue analysis map to help them evaluate the impact of the issue and develop a strategy that takes the issue into account. According to Professor King, "When each participant in the planning process is forced to deal with each issue in the specific terms of the strategic issue analysis model rather than in vague and undefined terms, real consensus is possible."

As a recap, here is a summary of the six procedures to follow to use the strategic issue analysis map.

1. Management identifies strategic issues.

2. Management and staff collaborate to define each issue formally.

3. Analysts build a model of the issue that identifies relevant subissues.

4. Management reviews the model and refines as necessary.

5. Staff gathers information on each subissue to build an "information model" for management.

6. Management uses the information from the strategic issue analysis map to aid in decision making.

Tool 63

Strategic Issue Analysis Map

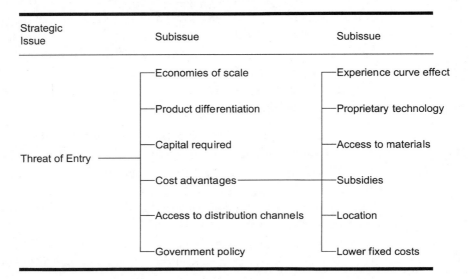

Strategic Issue	Subissue	Subissue
Threat of Entry	Economies of scale	Experience curve effect
	Product differentiation	Proprietary technology
	Capital required	Access to materials
	Cost advantages	Subsidies
	Access to distribution channels	Location
	Government policy	Lower fixed costs

Tool 64. Advantages/Disadvantages Table

What are the pros and cons of a proposed deal? Everyone has asked this question at least once, for it is probably the oldest form of business analysis. An advantages/disadvantages table formalizes the analysis by laying out pros and cons of a particular choice in an orderly format.

The simplicity of the method is deceptive; it is one of the best ways to think through a difficult issue. Not only is it an excellent method for presenting alternatives in a meeting or memo, but it is also a useful first step in thinking through almost any unfamiliar proposal and is especially well suited to strategic planning, where the choices may be unfamiliar and a subjective approach required.

Idea: Next time you are in a working meeting, try taking to the board with the advantages/disadvantages table as your scaffolding. With a little practice, you will become adept at defining pluses and minuses and will learn to use the table as a leadership tool.

Here are the instructions to follow to use the advantages/disadvantages table.

Lay out a table with two lines forming a "T" to create one column for advantages and a second for disadvantages. Start filling in whichever column is easier. If you favor a proposal, then you probably will find it easier to list several advantages. If you dislike the proposal, then the disadvantages will be obvious to you. Once one column has some entries, it is easier to fill in the other.

Tip: Groups usually generate more pros and cons than do individuals.

Challenge: After the table has been created, you must decide whether the project is worthwhile. You may need to study some of the issues before the choice can be made, especially if pros and cons are closely balanced. Also, when reviewing the list of advantages vs. the list of disadvantages, it is not advisable to make your decision based on a simple count of the number of items in each column. Often one or more advantages (or disadvantages) will outweigh the others in terms of relative importance.

In our example (see Tool 64), the table is being used to evaluate a proposed retail site for a fresh-baked cookie company. In this case, the advantages are compelling, and management will probably go ahead. The list of disadvantages should be used as an agenda; for example, someone needs to work on the security problems and negotiate storefront design.

Use the advantages/disadvantages table to:

- Decide whether to pursue an opportunity.

- Identify all the pros and cons of an alternative.

Tool 64

Advantages/Disadvantages Table

Advantages	Disadvantages
High-traffic area	Rent is high
No other cookie stores in the mall	Will not give an exclusive
Location is near movie theater	Restriction on storefront design
Parking available	Possible security problems at night
Plumbing included	

Tool 65. Quantitative Ranking Table

The quantitative ranking table method is a quick and easy way to formalize comparison of multiple alternatives. It is useful whenever many alternatives or criteria make the decision too confusing to work out in one's head. It is also especially helpful when a group cannot agree on which of several alternatives to pursue since it can be used to force agreement on which criteria to use and how to weight them. Once the decision criteria are established, the choice is more obvious and the group usually will reach consensus. (See Tool 65.)

Use this method when you have a set of well-defined alternatives. First, decide which criteria should be used to evaluate the alternatives. List all the criteria across the top of the table and list the alternatives down the left side. Now decide how important each criterion is to the selection and assign an importance rating to each. Use a 1-to-10 scale, where 1 equals unimportant and 10 equals most important.

Once the criteria and their weightings have been decided on, evaluate each alternative on the basis of each criterion. If a group is working on the decision, this step will require considerable discussion. Rate each alternative according to its strength on each criterion, using a 1 equals very weak and 10 equals very strong scale.

When all of the alternatives have been rated on each criterion and the table is full, discount these ratings to reflect the importance of each criterion. Divide each criterion rating by 10, then multiply this fraction times each of the alternative ratings in the criteria's column.

Tool 65

Quantitative Ranking Table

Alternatives	Cost (8)	Criteria (importance) Risk (10)	Time (5)	Resources (7)	Weighted score
A	4	8	6	5	17.7
B	7	9	8	6	22.8
C	3	6	10	5	16.9

Highest weighted score, Alternative B, is best.

Example: A criterion is given a 5 in importance, and an alternative is given a 10 on this criterion (5/10 = 0.5; 0.5 x 10 = 5). Therefore, the discounted score for this alternative is 5 for the criterion. After every score in the table has been adjusted in the same manner, sum the adjusted totals for each alternative and write the totals to the right. For Alternative A, the equation is (8/10 x 4) + (10/10 x 8) + (5/10 x 6) + (7/10 x 5) = 3.2 + 8 + 3 + 3.5 = 17.7. Rank the alternatives based on their scores; in general, the highest-scoring alternative is selected.

Use a quantitative ranking table to:

• Evaluate alternatives on the basis of multiple factors.

• Formalize the choice of alternative options or strategies.

• Help a group reach a consensus opinion.

Tool 66. Research and Development Matrix

Many books on research and development (R&D) offer checklists of issues for evaluating new product ideas. But Milton Rosenau of Rosenau Consulting (a Santa Monica firm specializing in technology management) finds these lists overly general and simplistic. Different issues are important for different products and companies. The problem: Which issues deserve the most attention in any specific case?

Here are the three instructions to follow to use an R&D matrix.

1. Develop a list of issues for evaluating a research proposal or new-product proposal. The objective is to list every issue that might be important to the future success of the project. Use one of the many published lists as a source, or compile your own by brainstorming with your managers or product development team to identify issues of relevance to your company.

    Useful device: Start with a list of general issues based on the functional divisions of your company. This list might include issues such as financing, development, manufacturing, distribution, marketing, sales, service, and social impact. Then break each area down into the specific issues that seem relevant to this project. For example, brainstorming on the marketing area might produce the following list: product life cycle, advertising, packaging, pricing, image, and market needs.

    An advantage of this approach is that you can delegate the task of listing relevant subissues to the managers specializing in each relevant area. The marketing manager, for example, could easily compile a list of all the relevant criteria under the marketing heading.

2. Rank evaluation issues on two scales.

- How *important* is the issue to success of the new product?

- How *strong* is the new product on this issue, based on its characteristics and the company's resources?

Use a numerical scale (0 to 5 or 0 to 10) for the ratings, with the high end of the scale equal to very important or very strong. (**Idea:** Have as many managers as possible rate the issues and calculate the mean ratings.)

3. Graph the results to summarize the findings.

The issues that fall into the lower-right quadrant of the graph are especially important to the project in the view of those managers polled. (See Tool 66.) In addition, they are issues on which the company and/or product is currently weak, and they obviously deserve special attention. Any project plan should discuss each of these in depth. Their impact on success should be evaluated carefully, and strategies to overcome weaknesses on these issues should be developed.

Note: The graph is useful in summerizing results to managers. But if many issues are rated, then the graph may become too cluttered. Try using a PC-based spreadsheet program to sort the list of issues by their two scores. Do a primary sort from highest to lowest of importance and a secondary sort from lowest to highest in strength. The most critical issues will be at the top of this list.

Caution: Rosenau warns that no checklist can substitute for judgment when evaluating R&D proposals. At best, a checklist is a useful aid to good judgment.

Use the R&D matrix to:

- Identify critical issues for planning or evaluating R&D proposals.

- Incorporate management judgment into evaluation of new products and new research projects.

- Present evaluation criteria in a graphical format.

As a recap, here are the three procedures to follow to use the R&D matrix.

1. Develop a list of issues that may be important for evaluating a proposed product or research project.

Research and Development Matrix

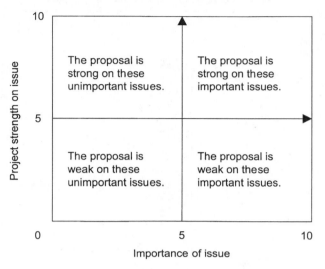

2. Rank each issue based on its importance to the success of the project and the company's strength in the area.

3. Plot the issues on a graph or sort them in a spreadsheet program to identify critical issues—those important issues on which the project is weak—and allocate resources accordingly.

Tool 67. Check Sheet

A check sheet is a simple data collection form consisting of multiple categories with definitions. (See Tool 67.) Data is entered on the form with a simple tally mark each time one of the events occurs. The most straightforward check sheet is simply a list of items that are expected to appear in a process. Users mark a check beside each item when it does appear. This type of data collection can be used for almost anything, from checking off the occurrence of particular types of defects to the counting of expected items (e.g., the number of times the telephone rings before answered) to making sure routine maintenance is performed (e.g., hourly cleaning of a mall's rest room facilities).

Here are the instructions to follow to construct a check sheet.

Clearly define the object of the data collection. Determine other information about the source of the data that should be recorded, such as shift, date, or machine. Determine and define all categories of data to be collected. Determine the time period for data collection and who will collect

Tool 67

Check Sheet

DATE: 5/8/00–5/14/00								
FLOOR: 4 EAST						Tray Delivery Process		
SHIFT: 7–3								
CHECK SHEET	MON	TUE	WED	THU	FRI	SAT	SUN	TOTAL
Tray Disabled				I	I	I		3
Production Sheet Inaccurate	I		I			I		3
Menu Incorrect	I		I		II		I	5
Diet Order Changed	I				I		I	3
Wrong Order Delivered	I		I		I	I		4
Patient Asleep	III	I	IIIII	I	II	III	I⊦	16
Patient Out of Room	IIIII	III	II	I	I	II	II	16
Doctor Making Rounds		II	IIIII	I	I	II	I	11
Patient Not Hungry	I	I	II	I	III	I	IIIII	14
Cart Faulty					I	I	I	3
Plate Warmer Broken	II	II	III	II	IIII	I	II	16
Heating Unit Broken	I	I			I		I	4
Thermometer Miscalibrated	I					I		2
Nursing Unavailable	III	II	III	IIII	II	III	III	20
Cart Unavailable	II	III	II		II		II	11
Elevator Malfunction			II	III		I		6
TOTAL	22	15	26	14	22	18	20	137

the data. Determine how instructions will be given to those involved in data collection. Design a check sheet by listing categories to be counted. Pilot test the check sheet to determine ease of use and reliability of results. Modify the check sheet based on results of the pilot test.

Tips:

- Use fishbone or Ishikawa diagrams (see Tool 35) or brainstorming to determine categories to be used on the check sheet.

- Construct an operational definition of each category to ensure data collected is consistent.

- Make the check sheet as clear, accessible, and easy to use as possible.

- Spend adequate time explaining the objective of data collection to those involved in recording the data to ensure the data will be reliable.

- Data collected in this format facilitates easy Pareto analysis (included as Tool 13).

Tool 68. Account Matrix

An account matrix breaks down a heading into the appropriate subcategory or subcategories. For example, during data entry into the general ledger, finance professionals determine expense allocation. Nonfinancial

Tool 68

Account Matrix

CLASSIFICATION OF NEW PRODUCT START-UP COSTS				
13 Types of Start-up Costs	Description	Product Cost	Period Expense	Capital
ACCEPTANCE & QUALIFICATION	Related to acceptance of the equipment; is equipment built to specification?			X
PILOT RUN	Testing of product in a plant environment.	X		
DIRECT LABOR COST PRIOR TO PRODUCTION	Individuals hired to perform direct labor functions prior to the start of production (usually in training or working on the pilot run).	X		
SPARE PARTS/TOOLING	Short-term life, prior to production, part of making equipment ready for production; or long-term life.			X
TRAVEL	Travel prior to start-up and related to the project.	X		
TRAINING	Costs of training employees on equipment, quality, etc.	X		
PLANT REARRANGEMENT	Physical changes to the plant such as walls and wiring; movement of equipment or fixtures.			X
INDIRECT SALARIES	Facilitators, supervisors, and engineers who devote 100% of time to the project prior to production.		X	
TECH. SUPPORT	Temp. engineers and consultants hired for project.	X		
REWORK	Pilot run production that is reworked prior to production start-up.		X	
PROD. CORRECTION	Rework due to changes in the specifications.		X	
AIR FREIGHT	A. Equipment			X
	B. Materials for pilot production		X	
RAMP UP VARIANCE	Caused by operating at less-than-standard load, at the start of production; short duration (8–12 weeks).	X		

managers need expense allocation information without being over-whelmed by detail. A plant controller can use an account matrix (see Tool 68) when explaining to managers at a manufacturing site the associated start-up costs for launching a new product. The matrix classifies the types of start-up costs that are expected, describes each one briefly, and shows how each type of cost will be recorded, that is, as product, period, or capital cost. In this example, the accountant has set up three distinct categories for allocating costs in language plant personnel can understand.

Tool 69. Grants of Authority Chart

To remain competitive, a company must achieve the maximum benefit from its fixed assets. Toward that end, the capital allocation function is vital. At the heart of an effective decision-making process regarding capital investments are procedures that are established and clearly documented. These procedures may include policies, capitalization guidelines, lease vs. buy instructions, and reporting requirements.

Companies often consider prospective capital projects during their annual budgeting process. These assessments facilitate the capital approval process when the actual requests are received. At many companies, proposals for capital projects are communicated to management via a formal request for authorization. These requests generally include focused statements reflecting the strategic direction supported by a wide assortment of research results, reports, and schedules justifying the project.

Often capital appropriation requests are approved at different management levels, depending on the amount of each request, and approval signatures must be obtained before a company makes commitments to spend money. The sign-off process also helps to assign appropriate accountability for the eventual success or failure of a project. An effective capital allocation procedure requires that everyone in the business understands and adheres to the amount of authority that he or she has been granted. Spending limits are a critical element of that authority. The grants of authority chart is one way to present spending limits. (See Tool 69.)

As part of a review of the capital allocation process, a company may want to examine its current spending limits and make revisions that better reflect changes in the organization's structure or an individual's responsibilities. To determine the appropriate limits, a company can evaluate the number, dollar value, and types of projects requiring approval at various levels. A grants of authority chart facilitates this process. For

Tool 69

Grants of Authority Chart

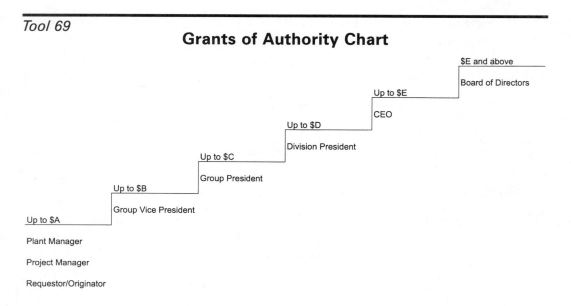

$E and above

Board of Directors

Up to $E

CEO

Up to $D

Division President

Up to $C

Group President

Up to $B

Group Vice President

Up to $A

Plant Manager

Project Manager

Requestor/Originator

example, one outcome of this type of analysis might be to triple (or to cut in half) various levels of spending authority.

Tool 70. Forecast Table

A forecast table compares various predictions of the future with what actually occurs. The format is appropriate for sets of highly labeled numbers. It shows at a glance the winners and losers in the forecasting game.

As you can see in this forecast table (see Tool 70), eight forecasters were asked for their predictions on key economic indicators. (Forecasters are not listed in categories for which they did not make a prediction.) The information for each indicator is listed in a separate column. The table depicts how the forecasts stack up against actual results (shown in the shaded row). For example, Wharton Econometric Forecasting predicted that Corporate Profits Growth would be +21 percent, and, actual Corporate Profits Growth turned out to be +13.3 percent.

The forecasts are sorted in *ascending* order with the *highest* forecast listed at the *top* of each column and the *lowest* forecast listed at the *bottom* of each column. The more accurate forecasts are listed in each column closer to the actual results. Forecasts that are too high are listed above the actual results, and forecasts that are too low are listed below the actual results.

Tool 70

Forecast Table

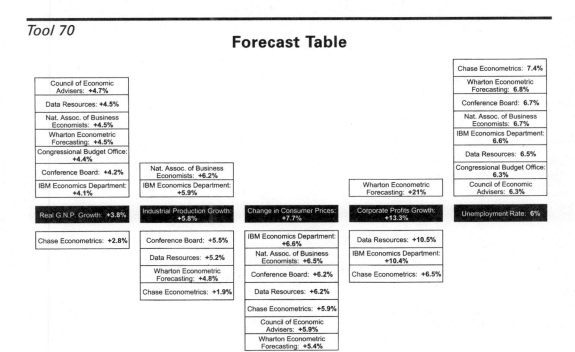

Tool 71. Food Guide Pyramid

Most Americans have seen the food guide pyramid. It depicts the six basic food groups and the number of recommended daily servings for each group. When it is printed on food containers, its message is relayed to a broad spectrum of the population, from rich to poor and from young to old. This is a simple and effective chart to get an important message across.

Consider Tool 71A; it contains the same data as the food guide pyramid, but it is presented here in the form of a histogram chart (included as Tool 13). Although the data is accurate and this is an effective chart, it is not a particularly powerful way to convey the message. It can be improved on, as depicted in Tool 71B, the food guide pyramid. Now it is a finished product that has power. As you can see, the image of the pyramid is timeless, timely, and solid. It is a clear path across the communications landscape.

A business application of the food guide pyramid is the salary guide pyramid. (See Tool 71C.) In this example, the human resources department of a company could use the design principles of the food guide pyramid in order to depict the components of employees' salaries.

Tool 71A

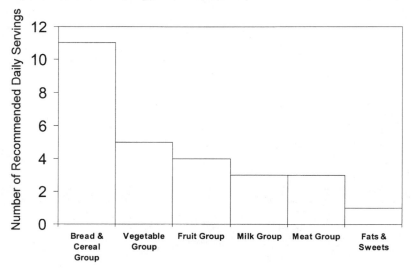

Food Guide Histogram

Food Guide Pyramid

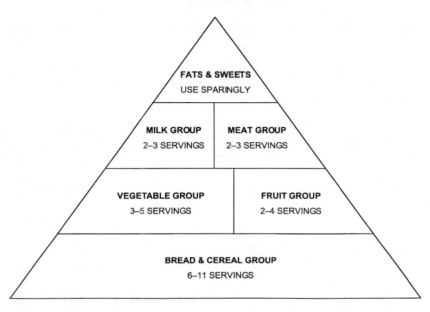

FATS & SWEETS
USE SPARINGLY

MILK GROUP
2–3 SERVINGS

MEAT GROUP
2–3 SERVINGS

VEGETABLE GROUP
3–5 SERVINGS

FRUIT GROUP
2–4 SERVINGS

BREAD & CEREAL GROUP
6–11 SERVINGS

Salary Guide Pyramid

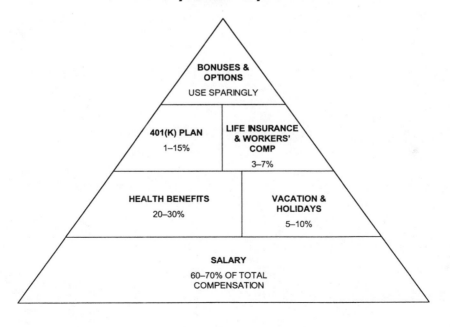

**BONUSES &
OPTIONS**
USE SPARINGLY

401(K) PLAN
1–15%

**LIFE INSURANCE
& WORKERS'
COMP**
3–7%

HEALTH BENEFITS
20–30%

**VACATION &
HOLIDAYS**
5–10%

SALARY
60–70% OF TOTAL
COMPENSATION

10. Planning, Scheduling, and Project Management

This chapter covers time charts, including time lines, PERT charts, and Gantt charts.

Tool 72. Time Line

A time line is a one-axis chart used to display past and/or future events, activities, or requirements, in the order they occurred or are expected to occur. (See Tool 72.) The major function of time lines is to consolidate and graphically display time-related information for purposes of analysis and communication. In addition to enabling the viewer to see graphically when things occurred or are to occur, a time line lets the viewer assess the time intervals between events. For example, in addition to seeing when each customer placed orders, one can see whether there is a pattern to the intervals between orders. If the chart is historical in nature and covers very long periods of time, it is sometimes referred to as a chronology chart. The axis of a time line can run horizontally or vertically. Time almost always progresses from left to right when the axis is horizontal. On a vertical axis, time might progress up or down. Progression from top to bottom is frequently used.

Tool 72

Time Line

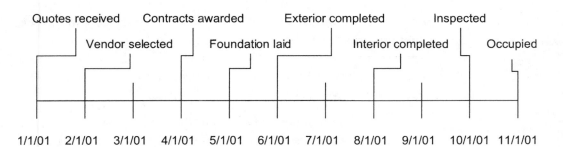

Tool 73. PERT Chart

PERT charts are time and activity networks that represent the major events and activities of a large program and show their interrelationships. PERT stands for *Program Evaluation and Review Technique*. The PERT system was introduced in 1958 with the Navy Polaris program. PERT charts are used to plan, analyze, and monitor programs. The advantage of PERT charts is that they provide clear pictures of whole programs. Among other things, they are helpful in determining how programs can be shortened and identifying which of the many subprograms are the most critical in assuring the overall program is completed on time.

A PERT chart uses a separate symbol for each event. An event may be a review meeting, a decision point, or simply the time at which one activity is completed and another begins. Lines or arrows then are used to connect the events in sequential fashion. The main body of a PERT chart generally starts and ends with a single node. The arrows and nodes are arranged in the same order they actually occur. The first event is represented by the first symbol on the left, the last event by the last symbol on the right; all others appear in their proper sequence in between.

The following three steps summarize the methods of using a PERT chart.

1. Prepare a network as the model of a carefully developed plan of action related to the program. Select specific events as milestones to be attained and represent them with a circle or an oval. An activity proceeds from one event to the other, from one event to several others, or from several events to one. After the plan for accomplishing the project as a whole is available, then the network is prepared as shown in Tool 73A, without considering the time element. The length of the lines used in the network have no meaning.

2. Estimate the time needed after a careful review of the completed network. The interrelationship of the events may be modified or subnetworks may be added if the selection of more events appears to be advisable. The time required to complete the project is based on estimates of the time needed to perform each activity between two events.

3. Identify the critical path to estimate the completion date. In Tool 73B, it will take 19 weeks to complete the activities that go from event A through events B, E, and F to H. This, in terms of expected time, is the longest path, which is called the *critical path* as compared to others, the *slack paths*. If the expected completion date

does not meet the requested date, then a reduction of the critical path has to be considered. This can be done by transferring some manpower from the slack paths or by replanning the network to eliminate certain tasks or perform more activities in parallel. The great advantages of using the PERT chart are that, at this point, the probability of meeting the required delivery date can be evaluated and the network focuses management attention on those areas where corrective action is most needed.

The bases of multiple arrows often meet at the right side of a node, indicating that the activity(s) on the left side of the node must be completed before any of the activities on the right side can be started. In Tool 73C, for example, the foundation must be laid before the wiring can be installed, the outer walls can be put up, or the plumbing can be completed. In PERT charts, it is not uncommon to have dozens or even hundreds of events and activities. A word description of each can be placed on or alongside the arrows or nodes.

Tool 74. Scheduling Chart

A scheduling chart is a chart that contains information about activities and dates. It helps you plan, manage, and communicate information about a schedule. You might use a scheduling chart to:

- Produce and maintain plans for a project.
- Schedule vacations or training sessions for a department.

Tool 73A

PERT Chart before Timing

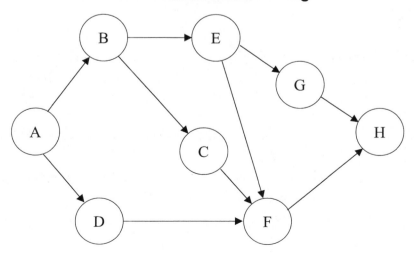

Tool 73B

PERT Chart after Timing

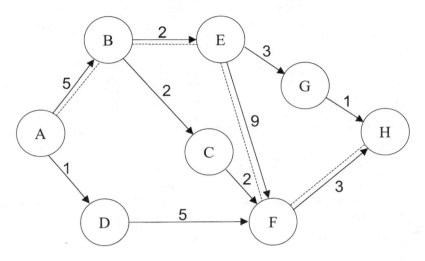

Tool 73C

PERT Chart

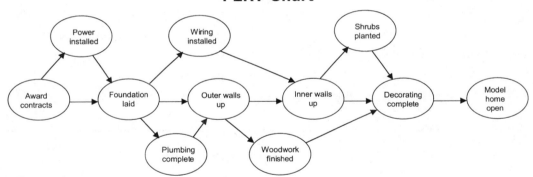

- See where backup is needed.
- Delegate activities to those covering for absent workers.
- Communicate assignments.

There are many different types of scheduling charts. Most have several things in common.

- The chart has a tabular format using rows and columns.
- The things being scheduled are shown on the vertical axis.
- Time generally is displayed on the horizontal axis.

Tool 74

Scheduling Chart

Vacation Schedule									
Week	**1**	**2**	**3**	**4**	**5**	**6**	**7**	**8**	**9**
Axson, D.		■	■						
Cram, D.								■	
Huff, S.				■	■				
Kritz, A.						■	■		
LeSage, D.									
Malkiel, B.	■	■							
Ribb, T.				■	■				
Scott, H.									■

- When something is scheduled, an entry is made in the cell(s) corresponding to the item scheduled and the appropriate time.

An example of a scheduling chart is shown as Tool 74.

Tool 75. Gantt Chart

A Gantt chart is a type of scheduling chart that documents the schedules, events, activities, and responsibilities necessary to complete a project or implement a group's proposed solution. It shows the steps involved in a project and their relationship over time. The duration of each step is indicated by the length of a horizontal bar. A Gantt chart is also known as a time and activity chart. There are many variations. Gantt charts are effective ways to keep a project on schedule by displaying a time line for each task.

A Gantt chart is actually a stacked bar graph. Excel does not offer a Gantt chart as an option, but you can produce one by using a stacked bar graph. In the Gantt chart represented in Tool 75, 12 tasks make up the project. Each task has a start date, a duration, and an end date. The chart tells you at a glance the critical path that will lead to the successful completion of the project and when tasks overlap.

Although there are many variations, all Gantt charts document the task or activity to be accomplished and the time period to perform the task. They allow a group to document the assumptions of its implementation plan. For example, if the plan is based on having a prototype built by February 6, then that assumption can be noted. The group then can develop contingency plans in case the deadline slips.

Tool 75

Gantt Chart

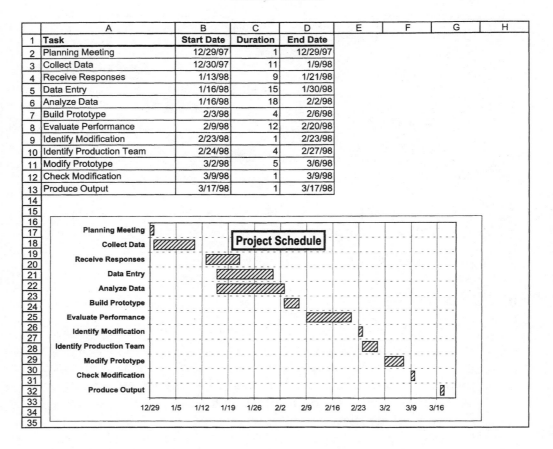

	A	B	C	D	E	F	G	H
1	Task	Start Date	Duration	End Date				
2	Planning Meeting	12/29/97	1	12/29/97				
3	Collect Data	12/30/97	11	1/9/98				
4	Receive Responses	1/13/98	9	1/21/98				
5	Data Entry	1/16/98	15	1/30/98				
6	Analyze Data	1/16/98	18	2/2/98				
7	Build Prototype	2/3/98	4	2/6/98				
8	Evaluate Performance	2/9/98	12	2/20/98				
9	Identify Modification	2/23/98	1	2/23/98				
10	Identify Production Team	2/24/98	4	2/27/98				
11	Modify Prototype	3/2/98	5	3/6/98				
12	Check Modification	3/9/98	1	3/9/98				
13	Produce Output	3/17/98	1	3/17/98				

How to re-create this Gantt chart using Excel

- Start with a new workbook and enter your task data, as shown. Column A contains the task descriptions; column B, the start date for each task; and column C, the number of days to complete the task. Column D contains formulas that determine the completion date for each task. For example, the formula in cell D2 is = B2 + C2 − 1.

- Create a stacked horizontal bar chart from the data in range A2:C13. The Chart Wizard probably will guess these series incorrectly, so you will need to set the category axis labels and data series manually. The category (X-axis) labels should be range A2:A13; the Series 1 data, B2:B13; and the Series 2 data, C2:C13.

- Remove the chart's legend and adjust the chart's height (or change to a smaller font) so that all X-axis labels are visible.

- In the Format Axis dialog box, select the following Scale options for the X axis: Categories in reverse order and Value (Y) axis crosses at Maximum category. This displays the tasks in order from top to bottom.

- Access the Format Axis dialog box for the Y axis. Set the Minimum and Maximum values to correspond to the earliest and latest dates in your project. Note that you can enter actual dates into this dialog box. To display weekly intervals, set Minimum to a Monday, Maximum to a Sunday, and Major Unit to 7.

- Select the data series that corresponds to the data in column B and go to the Format Data Series dialog box. Set Border to None and Area to None. This hides the first data series—the start dates—making the chart resemble a Gantt chart.

- Apply other formatting as desired. For example, you can add grid lines and a title. Tool 75 shows the completed chart after some touch-up work.

If you adjust your project schedule, the chart will be updated automatically. If you use dates outside the original date range, then you will need to change the scaling for the Y axis.

11. Probability, Prediction, and What-If

This chapter covers decision trees, payoff tables, and bright-line charts.

Tool 76. Decision Tree

A decision tree is a graphic representation of alternative decisions or actions that might be taken, plus potential outcomes resulting from those decisions and actions. The ability to see options and estimated outcomes before decisions are made is one of the main advantages of decision diagrams. Such diagrams can be used by almost any type of organization, to make simple or complex decisions, on a qualitative or quantitative basis. These diagrams might be used to establish a pricing policy, decide the size of a new plant, determine which research projects to pursue, or decide which proposal to accept.

When designing a decision tree, you may assign probabilities to certain branches and/or incorporate monetary values. This type of diagram is used sometimes when making risk decisions in which probabilities are assigned to each uncontrollable event. Such diagrams are typically horizontal, constructed from left to right, and are analyzed from right to left. (See Tool 76A.) For example, in analyzing a diagram, the most desirable outcome is selected from all possible outcomes (as listed in the far-right column). Next, the path is followed from right to left to see what decisions have to be made to achieve the desired outcome.

This method is helpful in planning a project or making a series of related decisions. It is used to look ahead at the various possible outcomes and map the specific decisions needed and their impact on outcomes. It relies heavily on management judgment, providing structure to the normal judgmental decision-making process. The method is easily used and, once mastered, is a useful back-of-the-envelope technique. It is also useful in presenting a project or problem to a group in a structured manner to get feedback on a decision or the assumptions on which the decision is based.

Follow these four steps when designing a decision tree.

1. Identify *all* of the feasible alternatives—the different decision options available to you. Also identify the different situations that

might prevail after the decision, focusing on a variable (or several) of relevance to the ultimate outcome of your decision.

Example: You must decide how much to spend on a new product launch. The marketing department presents three alternatives: low, medium, and high advertising and sales support. These are your feasible alternatives. You see two likely situations after introducing the product: (1) no competition and high demand or (2) a similar product introduction by a competitor and lower demand for your new product.

2. Draw a decision tree illustrating the decision alternatives and the situations. In Tool 76A, each decision is illustrated with a square, and each situation, with a circle. Lines connect the decisions and situations and are labeled to explain the diagram.

3. To calculate the possible financial returns from each combination of decisions and situations, enter the costs of each decision on the diagram (negative numbers) along with the incomes (positive numbers) from each situation and total them to calculate the payoffs. Write the payoffs at the right side of the diagram. (See Tool 76B.)

4. If you want, you then can refine the tree by estimating the probability of each situation occurring. (Probabilities should sum to 1.) Then multiply the payoff by the probability to calculate the "expected monetary value" of each alternative. Select the decision that will maximize expected monetary value. (See Tool 76C.)

Use decision trees to:

• Identify the options and potential outcomes of a decision or series of related decisions.

• Assign probabilities to events and calculate the likely returns from alternative decisions.

• Structure the decision task to identify where and how research should be used.

As a recap, here are the four procedures to follow when designing a decision tree.

1. Identify decision alternatives and alternative situations or "states of nature."

2. Diagram them on a decision tree.

3. Calculate the possible financial returns and costs and enter these on the diagram. Total to find the payoffs for each combination of decisions and situations.

4. *Optional:* Estimate the probability of each situation and weight payoffs by probabilities to calculate expected monetary values of each alternative. Select the decision with the highest expected value.

Tool 77. Payoff Table

A payoff table is one way of evaluating a choice in an uncertain environment. For example, a construction firm with a project in a foreign country might have to decide whether to hedge a currency when facing a potential devaluation.

To illustrate this technique, consider a payoff table for Lord Jim, the title character in the novel by Joseph Conrad. (See Tool 77.) In *Lord Jim,* Jim is first mate on board a ship called the *Patna.* One night the *Patna,* with 800 pilgrims on board, steams over floating wreckage and severely damages its bulkhead.

Tool 76A

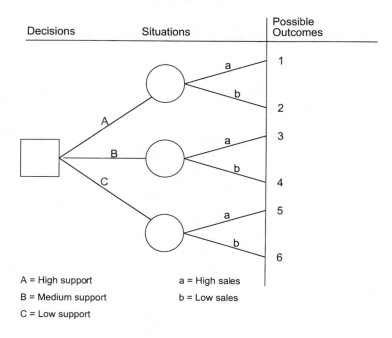

Decision Tree

A = High support
B = Medium support
C = Low support

a = High sales
b = Low sales

Tool 76B

Decision Tree—with Payoffs

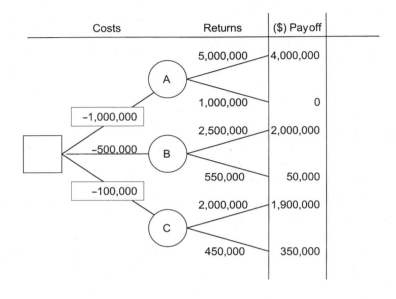

Costs	Returns	($) Payoff
	5,000,000	4,000,000
A		
	1,000,000	0
−1,000,000		
−500,000	2,500,000	2,000,000
B		
	550,000	50,000
−100,000	2,000,000	1,900,000
C		
	450,000	350,000

Tool 76C

Decision Tree—with Expected Monetary Value

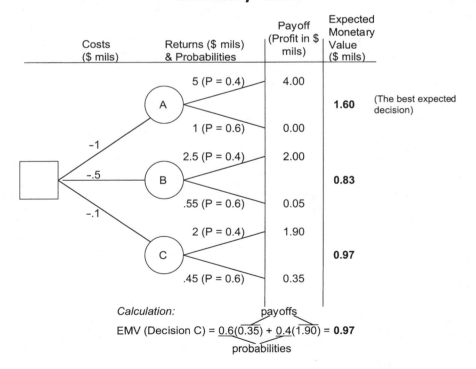

Costs ($ mils)	Returns ($ mils) & Probabilities	Payoff (Profit in $ mils)	Expected Monetary Value ($ mils)	
	5 (P = 0.4)	4.00		
A			1.60	(The best expected decision)
	1 (P = 0.6)	0.00		
−1				
−.5	2.5 (P = 0.4)	2.00		
B			0.83	
	.55 (P = 0.6)	0.05		
−.1				
	2 (P = 0.4)	1.90		
C			0.97	
	.45 (P = 0.6)	0.35		

Calculation: payoffs

EMV (Decision C) = 0.6(0.35) + 0.4(1.90) = **0.97**

probabilities

The crew (except for Jim) gets together, decides the boat is about to sink, and lowers the lifeboat. The entire crew is about to jump ship and leave the pilgrims to die. Jim is faced with the decision of either staying on board the *Patna* or getting into the lifeboat and rowing away. There are not enough lifeboats for everyone.

Jim has a choice: Jump or stay on board. Further, there are two possible scenarios not under Jim's control.

1. The *Patna* could stay afloat.

2. The *Patna* could sink.

To complete the payoff table, rank the various outcomes from best to worst for Jim.

- The best alternative would be if Jim stays on board and the boat floats. Jim would be a hero, and no one would die.

- The second-best alternative would be if Jim jumps and the boat sinks. Even though he would have committed a cowardly act, it would be justified. Because there were too few lifeboats and not enough time, jumping would be Jim's only salvation.

- The next-best alternative would be if Jim jumps and the boat floats. He would still be alive, but he would have committed a cowardly act, and he would live in disgrace because he deserted the ship.

- Finally, the worst alternative would be if Jim stays on board and dies at sea among the 800 screaming pilgrims.

The payoff table is a matrix. The rows are Jim's choices (jump or stay), and the columns are the ship's two scenarios (float or sink). There are four possible outcomes. The top row—Jim jumps—is shaded to emphasize that Jim could pursue a maxi-min strategy. By maximizing the minimum outcome, he can choose a strategy that lets him avoid the worst outcome. If he wants to avoid the worst outcome—his drowning—he can choose to jump and limit the outcomes to the upper row of alternatives. That way, regardless of what the ship does, Jim is not going to drown. He knows that he is going to live. The boat is either going to float or sink.

If Jim is risk averse, then he will pursue the outcomes that are available in the top row. However, if he decides to assume more risk, then he would choose to stay on board, making the lower row of alternatives the only possibilities. In this case, the only two possible outcomes for Jim are the best and the worst.

Just like Lord Jim at the helm of the *Patna*, a chief executive has to make the right decisions while piloting through uncharted waters.

Constructing a payoff table is a fast and systematic approach to evaluating choices in the face of uncertainty, when it takes courage to make a decision.

Tool 77

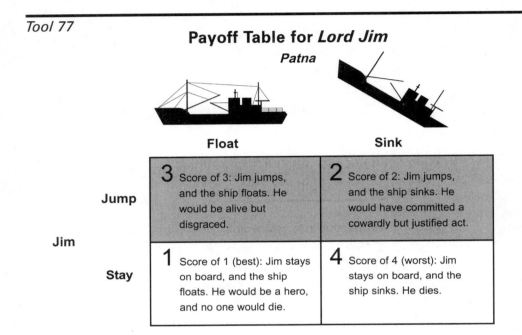

Payoff Table for *Lord Jim*

Patna

	Float	**Sink**
Jump	**3** Score of 3: Jim jumps, and the ship floats. He would be alive but disgraced.	**2** Score of 2: Jim jumps, and the ship sinks. He would have committed a cowardly but justified act.
Stay	**1** Score of 1 (best): Jim stays on board, and the ship floats. He would be a hero, and no one would die.	**4** Score of 4 (worst): Jim stays on board, and the ship sinks. He dies.

(Jim)

Tool 78. Bright-Line Chart

The bright-line chart is a welcome display when the topic for discussion revolves around "what-if" scenarios. A company might need a bright-line chart when, for example, it is developing its capital spending budget for the coming fiscal year.

Let us assume that a company's capital budget for the coming fiscal year is $75 million, and the company is organized into three operating divisions. Of the total $75 million, the Dedicated Systems Division has been allocated $30 million, the Telecommunications Division $25 million, and the Hardware Division $20 million. As you can see in Tool 78, the targeted amount of spending is depicted in the gray shaded row. ("Gray shaded row? Then why is it called a *bright-line* chart?" I hear you ask in surprise. Let me explain: The first time that I used this tool, I highlighted this row a bright yellow. During my presentation, we kept referring back to the "bright-line chart," hence the name has stuck ever since.)

The chart displays the three divisions' capital projects listed in columns. The name of each project is written in a separate rectangular box, together with the dollar amount of capital expenditure associated with the project. The heights of the boxes in each column vary in proportion to the size of the expenditures.

The projects are sorted in order of strategic importance, with the *most* important capital projects located at the *bottom* of each column and the *least* important capital projects located at the *top* of each column. All of the projects that each division plans to do, given the capital budget of $75 million, are listed below the bright line. These projects are the "must-do" projects for the coming year. For example, the Dedicated Systems Division plans to spend its $30 million on three projects: Project DS-101 (at a cost of $5 million), Project DS-102 ($9 million), and Project DS-103 ($16 million).

Listed above the bright line in each column are the potential projects that the divisions would "like to do" if additional capital resources became available. These projects have merit, but they are not essential to achieving their results for the year given the current constraints on capital spending.

By setting up the bright-line chart at time *T*, the company will already know what it will do should the $75 million capital budget change over the course of the year. If additional money becomes available at some point in the future (at time *T* + 1), and the capital spending level *increases*, then this would be depicted as the bright line being *raised*. When the bright line of $75 million is raised to $77 million at time *T* + 1, then everyone knows which capital project will be added first: In our example, the $2 million addition is allocated to the Telecommunications Division, and project TC-1 is the next project to get the go-ahead. This can be seen on the chart by the fact that project TC-1, which costs $2 million, is listed in the Telecommunications Division column immediately above the bright line. In a similar fashion, as additional money beyond $77 million becomes available, the bright line continues to rise. As the bright line rises, projects currently located above the bright line drop below the line, and they are thereby added to the "to-do" list for the year.

If, on the other hand, at time *T* + 1, the bright line is lowered to $73 million from $75 million, and the $2 million savings has to come from the Hardware Division, then project H-101 is history. This can be seen on the chart by the fact that project H-101, which costs $2 million, is listed in the Hardware Division column immediately below the bright line.

Uncertainty has always been a fundamental reason for developing a capital spending strategy. Yet, with the level of uncertainty that we see today, more and more people are asking "How can you develop a capital spending strategy in a world that keeps changing so fast?". They are afraid that articulating a rigid plan will hinder their ability to react quickly. I would argue that it is precisely at such times that you need a capital strategy. Think about what a capital strategy is: It is the process of making trade-offs and choices about how to allocate scarce capital resources. Only a company that has infinite capital resources does not need a capital strategy. Do you know of any company that has infinite resources?

This tool helps you to set capital budgeting priorities away from the heat of battle. Creating this chart involves a process of deciding what to do before action is required. It will guide you in advance of taking action. In the current environment, companies must have a plan with a clear set of strategic priorities in place. If you have a plan, then it is easier to make quick, confident decisions that are consistent with overall strategy. In the end, that arrangement allows for greater freedom, flexibility, and experimentation.

Tool 78

Bright-Line Chart

($ Millions)   **2001 Capital Expenditure**

	Telecommunications	Dedicated Systems	Hardware
$97	PROJECT TC-5 — 2		
	PROJECT TC-4 — 2		PROJECT H-2 — 4
	PROJECT TC-3 — 2		
	PROJECT TC-2 — 2	PROJECT DS-1 — 4	PROJECT H-1 — 4
	PROJECT TC-1 — 2		
$75	25	30	20
	PROJECT TC-101 — (5)	PROJECT DS-101 — (5)	PROJECT H-101 — (2)
	PROJECT TC-102 — (7)	PROJECT DS-102 — (9)	PROJECT H-102 — (7)
	PROJECT TC-103 — (13)	PROJECT DS-103 — (16)	PROJECT H-103 — (11)
$0			

12. Strategy

This chapter presents life cycle analysis, growth/share matrix, and value chain. It covers how to structure, test, and quantify strategies as well as how to visualize where you are competitively.

Tool 79. Life Cycle Matrix

The consulting firm Arthur D. Little (ADL) developed the life cycle matrix for use in portfolio planning and strategy development. It is based on the product life cycle model but adds a second dimension—competitive position. Because many factors contribute to the determination of both life cycle stage and competitive position, the matrix has considerable depth to it. It is especially useful for companies with product lines in rapidly evolving markets, including most technology-oriented companies, because product life cycles have such an impact on these companies. Consumer goods producers may find it a useful tool for the same reasons.

Follow these five steps to use the life cycle matrix.

1. Identify strategy centers, ADL's term for strategic business units, for analysis. Strategy centers are "natural businesses" having a well-defined market for their products or services and independent strategic objectives. The method can be applied to each of a company's strategy centers for a full portfolio analysis or to a single center to assess its strategic options.

   Note: ADL defines strategy centers as having products that have common competitors and customers; being affected similarly by changes in price, quality, and style; and being close substitutes for each other. Strategy centers also should have the potential to operate as stand-alone businesses if divested.

2. Next, identify the strategy center's industry maturity. The industry life cycle model postulates a five-stage cycle, moving from development, to introduction, through a growth stage, into maturity, and then followed by a decline stage. In practice, industries may stay in growth or maturity for a long time or may start to decline and then return to growth due to product innovation or changes in the environment. There are numerous references to the product life cycle graph in the marketing literature, and most managers are familiar with it (included as Tool 24). For the pur-

poses of Tool 79, we will combine the embryonic development stage with the introduction stage. Use Tool 79A as a guide to determining the appropriate stage. In selecting a stage, pick the one that best fits the strategy center, but do not be surprised if a center exhibits some characteristics that are inconsistent with this stage.

3. Define the current competitive position of the strategy unit. ADL defines competitive position as follows.

   Dominant: A strong leader. Rare. Technological leadership or partial monopoly usually necessary. An example might be Microsoft during the 1990s.

   Strong: Significantly greater share than any competitor and able to pursue strategies fairly independent of competitors. Not found in all industries.

   Favorable: Several competitors share leadership of their market, or one leads, but only by a little.

   Tenable: A weak number two or worse. A specialized niche competitor.

   Weak: Too small to survive in the long term given the nature of the industry. Or larger competitors that have fallen on hard times through serious mistakes or problems. In either case, needs to improve to survive.

4. Plot the strategy centers of your company on a table with the rows corresponding to competitive position categories and the columns corresponding to life cycle stages. In Tool 79B, we have simply written the names of the centers in the appropriate cells.

5. Interpret the matrix. It provides a simple way to visualize the portfolio in terms of the strength and life cycle stage of each business unit. A portfolio consisting of strategy centers that are in general strongly positioned but in mature and declining markets indicates a need to diversify into growth markets, for example.

Use the life cycle matrix to:

- Develop or evaluate strategies of business units.
- Manage a portfolio of businesses and product lines.

- Base strategies on life cycle stage and competitive position (especially relevant in high-tech or other industries where life cycles are short or volatile).

As a recap, here are the five procedures to follow when using the life cycle matrix.

1. Identify strategy centers—ADL's version of the strategic business unit.

2. Determine which stage of its product/industry life cycle each strategy center is in.

3. Determine which of ADL's competitive position categories applies to each strategy center.

4. Plot strategy centers on the matrix.

5. Interpret the matrix for developing strategy center strategies or for overall portfolio management.

Tool 79A

Life Cycle Matrix

Guide to Determining a Strategy Center's Industry Maturity Level

	INTRODUCTION	GROWTH	MATURITY	DECLINE
GROWTH RATE	Rising slowly	Accelerating	Leveling	Declining
SALES	Low	Rising	Peak	Declining
COSTS/ CUSTOMER	High	Average	Low	Low
PRODUCT LINE	Very short	Growing	Diversified	Shrinking
PROFITS	Negative	Increasing	Can be high	Declining
COMPETITORS	Few	Increasing	More but stable	Declining
TYPICAL PRICING	Cost-plus	Penetration	Competitive	Cut
ENTRY BARRIERS	Technology	Competitors	Competitors	Overcapacity
TYPICAL ADVERTISING	Awareness and education	Mass-market awareness	Differentiation and segmentation	Reduced

Tool 79B

Life Cycle Matrix

Competitive position:	Introduction	Growth	Maturity	Decline
Dominant				
Strong	Strategy Center A			
Favorable			Strategy Center B	
Tenable				
Weak	Strategy Center C			Strategy Center D

Tool 80. Growth/Share Matrix

The Boston Consulting Group (BCG) growth/share matrix is a framework for portfolio planning in a diversified company with many businesses. Each business is located on a two-dimensional grid. One dimension represents industry attractiveness, summarized by the real annual rate of market growth. The other dimension represents the business's competitive position, summarized by its market share relative to its largest competitor.

BCG developed the matrix to help diversified companies maximize portfolio performance by identifying which products to invest in, which to milk for investment funds, and which to eliminate from the portfolio. Portfolio planning models such as the BCG matrix are deliberately oversimplified to reduce the amount of data and suggest priorities for further analysis. Their overall thrust, however, is toward investing in high-share businesses in high-growth markets ("stars") and divesting low-share businesses in low-growth markets ("dogs"). Most profit and cash is generated by "cash cows" (high share, low growth). There is much debate about whether to continue investing in low-share businesses in high-growth markets ("question marks").

Tools 80A and 80B are growth/share matrixes. The vertical axis displays market growth rate, and the horizontal axis displays relative market share. Note that in these tools the relative market share *decreases* as we move from left to right. The size of each circle is directly proportional to its total sales. In Tool 80B, there is a circle for each type of product (stainless steel, super alloy, structural ceramics, and tool steel) manufactured by SBJ Technology. This analysis helps categorize the performance of the products plotted (i.e., stars, dogs, cash cows, or question marks).

The matrix assumes that cash-flow potential of a product is related to the overall growth rate of its market, to its share of the market relative to competing products, and to its current size. Since these are generalized measures, they allow comparison of widely differing products and markets.

The matrix depends on the following cash-flow assumptions. The experience curve (included as Tool 33, in which costs go down with number of units produced) justifies the assumption that high market share relative to competitors is advantageous. The product life cycle graph (included as Tool 24, in which a product moves from introduction through growth, maturity, and decline stages) justifies the assumption that high growth is desirable. Other arguments also favor high share (economies of scale and tactical strength of leadership position) and high market growth (higher sales growth and higher profits). These arguments lead to the cash-flow predictions in the growth/share matrix.

Here are the six steps to use the growth/share matrix.

1. Identify the unit of analysis. Usually the method is applied to individual products, and they are looked at in the context of their entire markets. But sometimes a product category is subdivided (e.g., models). Markets may also be subdivided (e.g., geographic areas, segments). The matrix also has been used to analyze strategic business units, divisions, entire companies, and even countries. As long as the cash-flow assumptions of the model appear valid, then the matrix may be used for whatever unit of analysis is important in decision making.

2. Gather the necessary statistics on the products to be analyzed.

   - Annual sales for each of your company's products (in dollars or units, most recent period)

   - Annual sales of the largest competitor to each of your products (same unit and period)

   - Annual growth rate in each of the product markets (percent growth in total market revenues or unit sales over most recent period)

3. Calculate relative market share by dividing most recent annual sales of each product by the annual sales of its largest competitor. Relative share is 1, where your company's product has as large a

share of its market as its main competitor. If your product has a smaller share than its largest competitor, then its relative share is less than 1. If it is larger than the next largest competitor, then its relative share is more than 1.

4. Plot each product on a graph with market-growth rate on the vertical and relative market share on the horizontal. A log scale usually is used for relative share, and relative share decreases to the right. The graph customarily is divided into quadrants by a vertical line at relative market share equals 1 and a horizontal line at market growth equals 10 percent. Draw a circle around each product's plot with a diameter proportionate to the product's recent annual revenues. (Some users make the *area* of the circle relative to revenues—a more accurate but complex approach.) *Hint:* Plot sales on a bar chart that is half as long as your growth/share matrix's horizontal axis. Trace the bars to obtain appropriate circle diameters.

5. Evaluate the product portfolio based on the quadrant each product falls into. Products are categorized according to the quadrant where they appear in the matrix. High-growth, high-relative-share products are stars. Low-growth, high-relative-share products are cash cows. Low-growth, low-share products are dogs, and high-growth but low-share products are question marks. The cash-flow assumptions of the model lead to the following conclusions about each of the quadrants.

Stars: High growth and high share. Strongly positioned in the growth phase of the product life cycle. Although cash flow is strong, it may not be sufficient to finance the rapid growth. Expected to throw off excess cash later when the market matures and growth slows.

Cash cows: Dominant products in a mature market produce excess cash that can be invested in stars or question marks.

Dogs: Low-share products in low-growth markets. Cash flows are low and often negative due to the weak competitive position. If investment is required to maintain the share of a dog, then it might be better to divest it and reallocate the funds to a question mark or star.

Question marks: Low share of a growing market suggests weak cash flow and considerable cash needed to maintain share.

Investment in a question mark may increase share and create a future star, but investment is risky.

Managers usually cannot control market growth but can invest to maintain, lose, or build market share. Market share strategies for each product should reflect the need for future cash cows. Unprofitable dogs can be dropped to divert funds to question marks or stars, and efforts can be concentrated on the most promising question marks. (Share strategies should produce a cash flow from the lower left to the upper right of the matrix.)

6. Follow the movement of products over time. A time series makes it possible to identify the direction of each product's movement within the growth/share matrix. Prepare several matrixes on the same scale, one for each of the last two or three years, and superimpose them. Trace them onto a single matrix or use directional arrows to indicate movement on a copy of the most recent year's matrix. According to Bruce Henderson, founder of BCG, matrixes "may be even more useful to evaluate management performance." The manager of a product or a portfolio of products can be evaluated based on how products move within the matrix over time. Any product that does not move rapidly toward 100 percent of the relevant market share is presumably mismanaged.

   Cautions: The growth/share matrix is not applicable where experience curve and economies of scale are irrelevant to product cost. It assumes units of analysis are (1) irrelevant to product cost and (2) independent. If resources are shared among business units, cash-flow data and assumptions should be questioned. How markets are defined can change the outcome of the analysis dramatically by altering relative share and market growth. This matrix is not useful where balancing cash flow is not an important objective (e.g., where external funding is planned).

Use the growth/share matrix to:

- Develop market share strategies for a portfolio of products based on their cash-flow characteristics.

- Represent a firm's product portfolio so as to highlight its strengths and weaknesses.

- Decide whether to continue investing in unprofitable products.

- Allocate a marketing budget among products so as to maximize long-term cash flow from them.

• Measure management performance based on the performance of a manager's products in the marketplace.

As a recap, here are the six procedures to follow when using the growth/share matrix.

1. Identify the unit of analysis, being careful to define the market realistically. The matrix can be used for products, business units, or other units of analysis.

2. Gather statistics on annual sales, competitors' annual sales, and annual market-growth rate for each product or business unit to be analyzed.

3. Calculate relative market share (the revenues of a unit divided by the revenues of its largest competitor).

4. Plot the product/business units on a graph using relative share and market-growth rate. Divide the graph into quadrants to create the growth/share matrix.

5. Evaluate the portfolio based on BCG's assumptions concerning cash flow within the matrix and performance of products/units in each quadrant.

6. Follow the movement of products over time.

Tool 80A

Growth/Share Matrix

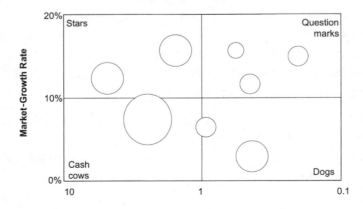

Relative Market Share

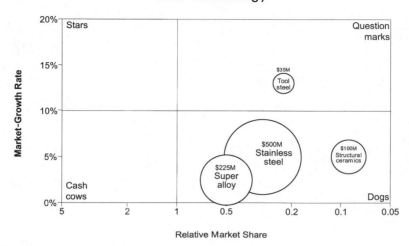

Growth/Share Matrix
SBJ Technology

Tool 81. Return on Sales/Relative Market Share Matrix

Tool 81, the return on sales/relative market share matrix, is similar to Tool 80, the growth/share matrix; only now the vertical axis contains a different metric for industry attractiveness. As the name of this tool implies, the vertical axis displays return on sales, and the horizontal axis displays relative market share. The sizes of the circles are proportional to total sales.

Tool 81 depicts the U.S. specialty steel market. This matrix is used in a similar fashion to Tool 80. The typical applications of this matrix are planning acquisitions and divestitures, developing market strategies, performing competitive analysis, and comparing financial performance.

Tool 82. Growth/Growth Matrix

Tool 82 is a growth/growth matrix. The vertical axis displays percent annual change in revenue, and the horizontal axis displays percent annual change in market share. The sizes of the circles are proportional to annual revenue. The application of this particular matrix is to analyze the market growth in California between 1996 and 2000; each circle represents a different city in California. In a business application, the growth/growth matrix would be used to evaluate potential markets and to suggest an action that might be taken (e.g., invest heavily, divest, analyze in more detail, put on fast track, or deemphasize).

Tool 81

Return on Sales/Relative Market Share
U.S. Specialty Steel Market

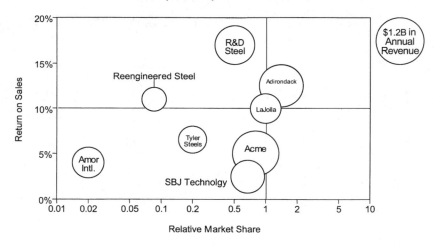

Tool 82

Growth/Growth Matrix
California Market Growth

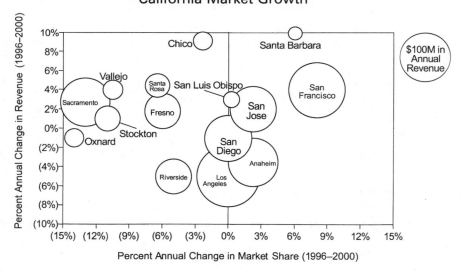

Tool 83. Customer Profitability Matrix

Companies often price on the basis of average product cost, but prices rarely reflect the total costs of providing the product or service to a *specific* customer. A study by Elliot Ross of McKinsey, in collaboration with

Harvard Business School faculty, indicates that the cost of servicing individual customers varies by as much as 30 percent, making some customers profitable to work with and others unprofitable. The study went on to identify types of customers based on their buying behavior and average profitability, making it possible to predict the profitability of each customer type and tailor strategy toward the most profitable customers.

There are many examples of how companies treat types of customers differently. A platinum-level frequent flyer member gets access to the VIP lounge at the airport and can board the plane first. The Land's End Company takes incoming calls from repeat customers before it takes incoming calls from first-time buyers. The customer profitability matrix helps companies analyze their customers and develop strategies that target the most profitable customers.

Tool 83 is one way to visualize customer profitability.

Here are the five steps to follow when using the customer profitability matrix.

1. The hitch is that you must first analyze costs per customer in more detail than most cost-accounting systems allow; you need to calculate total costs of servicing each customer over a period of time. There are different approaches to this problem, depending on how far you want to go with the method. To start, try working with a small, random sample of accounts to limit the amount of homework needed, or just perform the analysis on the national or "A" accounts. A more involved alternative is, of course, to hire an experienced consulting firm to do the analysis.

   Identify and allocate costs as thoroughly as possible in the following categories.

   - *Presale costs:* Differences in customer buying processes, location, and need for customized engineering and other presale services make a big difference in costs per customer.

   - *Production costs:* Some firms do custom designs for their customers, which obviously affects production costs per customer. But other factors—packaging requirements, timing, setup time, fast delivery requirements—also may vary by customer. Inventory and accounting procedures can make it difficult to estimate these costs on a per-customer basis.

   - *Distribution costs:* The customer location and the mode of shipment can vary significantly from customer to customer. The logistics or transportation departments usually can identify per-customer distribution costs easily, although they are rarely asked to.

Tool 83

Customer Profitability Matrix

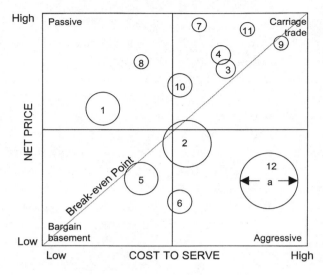

a = Total Annual Sales

CUSTOMERS 2, 5, 6, 9, AND 12 ARE UNPROFITABLE

- *Postsale service costs:* Training, installation, support, repair, and maintenance costs usually vary, and warranty and contract terms can differ as well, making the cost of postsale service an important variable for customer cost analysis.

2. Calculate the net price of your products or services to the same customers for which you performed the cost analysis. Make sure that any discounts, returns, and so forth are taken into account. In many companies, these items are not reported on weekly or monthly account-by-account revenue reports; they appear later as adjustments. Also collect a gross-sales-per-year figure for each customer analyzed.

3. Plot the data for each customer on a graph, with net price on the vertical axis and cost on the horizontal axis. (See Tool 83.) Represent each customer with a circle proportional to gross sales (or to revenues from that customer as a proportion of your company's total revenues). In our example, there are 12 customers, and each circle has been assigned a number 1 through 12. Add lines to represent average price and cost and a diagonal line to show the break-even point where cost equals price.

The resulting matrix shows which customers have high costs in relation to the price they pay and which customers are producing high net profits for your company. It divides customers into four groups based on where they stand relative to the average cost and price lines. Quadrants are labeled "carriage trade," "bargain basement," "aggressive," and "passive," reflecting the fact that there is a typical buying behavior for each quadrant of the matrix. Customers can be divided into these four groups according to price and cost levels.

Carriage trade: High cost, high net price. Willing to pay a high price for superior products and services. Often place small, custom orders.

Bargain basement: Low cost, low net price. Price-sensitive; less concerned about quality or service.

Aggressive: High cost, low net price. Command high quality and service and low prices, often because they buy in large quantity. Strong negotiators and technological leaders often fall into this category as well.

Passive: Low cost, high net price. Less concerned about quality or service but not very price-sensitive either.

4. Interpret the matrix and develop strategies and support systems to help your company manage its customers for profitability. Most companies have customers in all four categories. Sometimes a company's most prized accounts—large-volume, high-profile customers—are breaking even or losing money when a careful cost analysis is performed, while many of the smaller accounts may be far more profitable.

On the strategy level, a company can define itself based on the type of customer it pursues. For example, a company positioned in the high-price/high-quality end of the market should focus on customers in the upper-right section of the matrix, while low-price/low-service providers looking for high volume and a favorable cost position in their markets should target customers in the lower-left section of the matrix. Note that customer buying behavior may vary depending on who makes the purchase and what it is. Changes in customers' organization or operations also may move them from one quadrant to another.

Idea: Survey decision makers in your markets to identify company buying behaviors in terms of service needs, negotiation, and other variables related to net cost. Look for relationships between net cost of serving a customer and other variables, such as size, age, location, and organization, that will help you define and target profitable segments for account development.

On the tactical level, pricing should reflect per-customer costs to the extent that they can be estimated. Information systems may need to be modified to provide the information necessary for negotiation and discount decisions as well as for establishing base prices.

5. Repeat the analysis. Cost structures change over time and so do customers. For example, a reorganization may change the way a customer buys your product, altering the cost of serving that customer and requiring a change in your pricing tactics. Account profitability strategies, like any strategies, should be evaluated every year or two, and up-to-date information will be needed for this purpose.

Use the customer profitability matrix to:

- Allocate sales force effort based on account profitability.
- Develop sales and marketing strategies that target the most profitable types of customers in a market.
- Provide detailed cost data for customer price negotiations.
- Find out how profitable the largest, most demanding customers really are.

As a recap, here are the five procedures to follow in order to use the customer profitability matrix.

1. Analyze costs per customer, including presale costs, production costs, distribution costs, and postsale service costs.

2. Calculate the net price and total annual sales for each of the customers included in the cost analysis.

3. Graph the customers on a cost/price graph. (Use circles proportionate to total annual sales.) Add lines to represent average cost and price and include a diagonal break-even line. The matrix indicates which customers fall into standard categories of buyer behavior: carriage trade, bargain basement, aggressive, and passive.

4. Develop strategies and support systems to help the sales force and the marketing department target customers in the category or categories considered most desirable in your current situation.

5. Repeat the analysis periodically in order to identify changes in account profitability and renew the sales profitability strategy.

Tool 84. Attractiveness/Strength Matrix

Tool 84, the attractiveness/strength matrix, is an example of a business matrix where two variables are encoded by means of pie charts. A third variable is encoded by the size of the pie charts.

The attractiveness/strength matrix was developed at General Electric (GE) in the 1970s in response to problems they saw with the Boston Consulting Group (BCG) growth/share matrix (included as Tool 80). It provides a similar comparison of market attractiveness and business strength, but where the BCG matrix measures attractiveness by growth rate and strength by relative market share, the GE matrix uses a larger number of factors. Because it uses multiple factors, it can easily be adapted to the specific interests of management or the particulars of an industry by changing the factors and their emphasis.

The matrix is drawn by identifying critical internal and external factors and then weighting these to create measures of market attractiveness (external factors) and business strength (internal factors).

The following five instructions provide a step-by-step approach to this task. As with the BCG matrix, it is important to define the unit of analysis carefully before starting to collect data. Focus on meaningful strategic business units and define markets carefully.

1. Define factors. Select the important factors for evaluating business (or product) strength and market attractiveness. In GE's terminology, market attractiveness factors are "external" and business strength factors are "internal." Here is a list of factors to guide your selection. (You may want to drop some of these and add some of your own.)

Market Attractiveness/ External Factors	Business Strength/ Internal Factors
Cyclicality of sales	Advertising
Demographics	Breadth of product line
Entry barriers	Customer service
Environmental issues	Distribution

Exit barriers

Market concentration/
structure

Market-growth rate

Market size

Political issues

Profitability

Regulation

Resources availability

Social issues

Technological advances

Financial strength

Image

Management strength

Manufacturing

Market share

Marketing

New-product development

Perceived quality

Repair and support

Sales force

This step is very important: If you overlook important factors, then the analysis will not be valid. Do not include trivial factors; they will sidetrack the analysis and waste time. Use a management team to identify factors, perhaps by brainstorming.

2. Assess the impact of external and internal factors. Starting with the external factors, review the list (using the same group of managers) and rate each factor according to how attractive or unattractive it is. (*Tip:* For factors having a similar impact on all competitors, evaluate impact in general; but where a factor impacts competitors differently, compare the impact on your business with the impact on key competitors.)

Use a five-point rating scale where:

1 = very unattractive

2 = unattractive

3 = neutral impact

4 = attractive

5 = very attractive

Example: A highly cyclical pattern of sales might be ranked 2, unattractive.

Now, for the internal factors, do a similar rating using the following scale where:

1 = major competitive disadvantage

2 = competitive disadvantage

3 = equal to competitors

4 = competitive advantage

5 = major competitive advantage

In rating your business on the internal factors, generally you must compare your business to a competitor. Pick the strongest overall competitor as a basis for comparison rather than switching to the strongest on each specific factor, or else your ratings will be artificially low.

3. Assess the importance of external and internal factors and develop summary measures of strength and attractiveness. Now that you have established a rating for each factor, you need to decide how important each factor is to a general assessment of your business's position. There are two approaches to this task:

Qualitative assessment: Review and discuss the list of internal factors and their ratings. Judge how strong the business is on the basis of the factor ratings using the following scale: high, medium, and low. Now evaluate the attractiveness of the market based on the ratings of external factors. Decide whether market attractiveness is high, medium, or low.

Quantitative assessment: Apply the following instructions separately to the list of internal factors and the list of external factors, in two steps.

Rank the factors by importance, placing the most important at the top of the list. (There may be many ties.) Now assign a weight to each factor, using fractions that add to 1 or percentages that add to 100 percent. This is easier said than done. A helpful trick is to start by giving all factors the same rating, which is equal to *1 divided by the number of factors,* and then adjusting up or down to reflect rank. If you do this on a computerized spreadsheet showing the sum of ratings, then you can keep adjusting them until the total is correct.

Next, multiply each factor's ranking on the 1-to-5 scale (from step 2) by its importance weight. ***Example:*** Market share confers a competitive advantage on your firm relative to its primary competitor; the managers gave this internal factor a 4. It was considered one of the most important factors and was assigned a weight of .2. To calculate this factor's weighted score, multiply .2 by 4, which equals .8. Total the weighted scores for both the internal and external factors (keeping the two lists separate). The totals will be somewhere between 1 and 5, with 1 representing low market attractiveness or low business strength, and 5 representing high market attractiveness or high business strength.

4. Plot the business on the matrix. (See Tool 84.) Construct the matrix by labeling the horizontal axis of a graph "Market Attractiveness" and the vertical axis "Business Strength." Divide the graph into grids by dividing each axis in equal thirds with two lines. Label the grids high, medium, and low on each axis (high to the left and top). If you used the quantitative method of rating (step 3B), then superimpose a 1-to-5 scale on each axis.

   Now take the overall score of the business for each group of factors and plot it on the matrix. Plot the external factors on the horizontal market attractiveness axis and internal factors on the vertical business strength axis. Repeat the analysis for other business units or for competitors of the business unit you are studying, depending on the strategic interests of management.

   You can perform the analysis for several time periods to observe the movement of businesses within the matrix over time.

   Idea: When plotting a portfolio of business units, each in a different market or industry, represent each business as a circle with the diameter proportionate to the total sales in its market and draw a pie-shaped wedge within the circle to represent the business's market share. Otherwise, share information will not be visible on the matrix since it is combined with other internal factors in the business strength assessment.

5. Interpret the matrix. The following table, based on the thinking of the consulting firm A.T. Kearney, Inc., presents the strategic implications of the matrix for portfolio planning.

This method also can be used to forecast the market attractiveness and business strength of a portfolio of business units. Have a team of managers assess each factor, as above, except that they should try to anticipate the impact and importance of each factor at some point in the future. Then build up an overall score for strength and market attractiveness, as above, and plot a projected matrix.

Use the attractiveness/strength matrix to:

• Evaluate business units based on their strength in the market and the attractiveness of their markets.

Market Attractiveness	Business Strength	Suggested Strategies
High	High	Grow Seek dominance Maximize investment
Medium	High	Identify growth segments Invest strongly Maintain position elsewhere
Low	High	Maintain overall position Seek cash flow Invest at maintenance level
High	Medium	Evaluate potential for leadership via segmentation Identify weaknesses Build strengths
Medium	Medium	Identify growth segments Specialize Invest selectively
Low	Medium	Prune lines Minimize investment Position to divest
High	Low	Specialize Seek niches Consider acquisitions
Medium	Low	Specialize Seek niches Consider exit
Low	Low	Divest Focus on competitor's cash generators Time exit

- Represent a firm's business unit portfolio so as to highlight its strengths and weaknesses.

- Develop matrix-based strategic planning where broad and flexible definitions of market attractiveness and business strength are desired.

As a recap, here is the five-step procedure to follow when using this matrix.

1. Define the critical success factors you want to use in evaluating market attractiveness and business strength.

2. Assess the impact (positive, negative, or neutral) of internal factors on business strength and external factors on market attractiveness.

Tool 84

Attractiveness/Strength Matrix

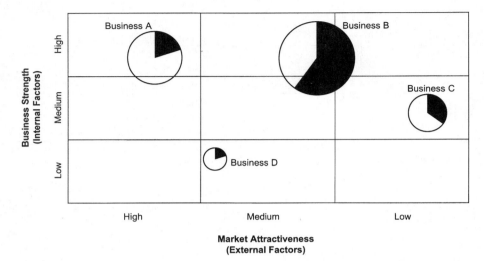

3. Assess the importance of factors and develop summary measures of strength and attractiveness.

4. Plot the business unit (usually with a portfolio of businesses) on the matrix.

5. Interpret the strategic implications of the matrix.

Tool 85. Strategic Planning Process Pyramid

The strategic planning process pyramid (see Tool 85) depicts a 10-step strategic planning process. This process starts when executives define a mission for the company (step 1); next, the company managers complete a situation analysis (step 2), and so on down the pyramid, until the process is complete.

At the bottom of the pyramid, rolling six-month marketing plans (step 10) form the base of the pyramid. These six-month forecasts flow from the budget (step 9); the budget, in turn, flows from the strategic plan (step 8). The first year of the three-year strategic plan becomes the budget, and the rolling forecast confirms that the budget will be achieved.

Other companies have different processes. An alternative approach is to develop a rolling six-quarter (18-month) forecast. A continuous forecast process can make a formal budgeting process unnecessary—one of the forecasts simply becomes the budget for the upcoming year. Other

Tool 85

Strategic Planning Process Pyramid

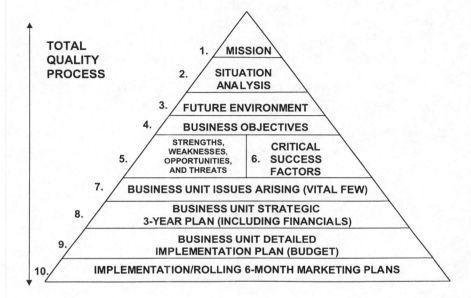

organizations forgo a continuous forecasting process altogether, relying instead on pulling together a revised outlook on an exception basis.

Whatever the strategic planning process, it can be depicted in a pyramid.

Tool 86. Value Chain

Michael Porter's book *Competitive Advantage: Creating and Sustaining Superior Performance* presents the concept of the value chain and shows how it can be used as the fundamental tool in diagnosing competitive advantage. Value chain analysis is a complex and rich device for strategic planning. It allows managers to separate the underlying activities a firm performs in designing, producing, marketing, and distributing its product or service. Use the value chain to understand the behavior of costs and how to create and sustain a competitive advantage. (See Tool 86A.)

The concept behind value chain analysis is simple. The activities of a company add value in excess, it is hoped, of the costs of those activities. Value chain analysis looks at where and how a firm adds value, breaking down activities and assigning a value added to them rather than assigning costs as in the conventional accounting system. The knowledge gained from value chain analysis is helpful in understanding how competitors differ and how to create competitive advantage.

Here are the five steps to follow when doing value chain analysis.

1. Select the unit of analysis. Use value chain analysis to study individual business units, competing in specific industries. Do not try to analyze an entire industry—the analysis should be competitor by competitor. And do not apply the method to all the activities of a firm if the firm competes in several discrete industries.

2. Identify primary value-adding activities within the chosen business unit. Primary activities are defined as follows.

Purchasing, inventory holding, and materials handling:	All inbound logistics including procurement, receiving, storing, and handling product inputs
Production:	Transforming raw materials and other inputs into the product and creating and maintaining the facilities needed to produce the product
Warehousing and distribution:	All outbound logistics involved in moving the product to buyers, including warehousing and distribution
Sales and marketing:	Bringing the product to buyers and inducing them to buy and use it
Customer service:	Installing, maintaining, and repairing the product; training, parts, and other services

These activities are generally quite distinct, having different economies and representing significant cost centers within the business unit. For each category of primary activity, prepare a list of the major component activities. For example, direct selling would be a component activity in the sales and marketing category.

Porter recognizes three types of activity, all of which can be found in most companys' primary activities. *Direct* activities are those activities that most obviously add value—machining is a direct activity under operations. *Indirect* activities are those that are necessary in order to perform the direct activities—overhead and administration included, for example, servicing machining equipment. *Quality awareness* activities are those that assure that the other activities are performed properly, such as testing machined parts for conformance to product specifications.

3. Identify support activities. Support activities are divided into the following three categories.

Human resources management and development:	Finding, training, and keeping the people needed to perform primary and support activities
Technology research, development, and design:	Research and product development in support of products or processes used to produce them
Infrastructure activities:	Management, planning, finance, accounting, and other activities often considered to be overhead

Look for the specific direct, indirect, and quality-assurance activities that make up each of these three categories. List the key activities by category, indicating their type.

4. Identify linkages in the value chain. Porter defines linkages as "relationships between the way one activity is performed and the cost or performance of another." For example, parts procurement and assembly are tightly linked in a just-in-time (JIT) production process. Linkages often are harder to identify than discrete activities—it takes insight into the business to see where the important linkages exist and to see opportunities to create new linkages. JIT production is in fact a good example of how a linkage can be exploited to create competitive advantage.

One way to find linkages is to look at information flows. In general, cooperation, and therefore communication, is needed between different functional groups within the business unit in order to exploit a linkage.

5. Study the value chains to identify sources of competitive advantage. Porter uses a graphic form for presenting value chains. Tool 86A shows a sample value chain diagram. Use this diagram, adapted to the business unit you are studying, to represent the activities you have identified. Allocate space within the diagram in proportion to the contribution of each activity to value added.

When two competitors' diagrams are compared, differences in primary or support activities are usually noticeable. Even though two companies may produce the same product or service for the same market, each will emphasize different activities in differing proportions. These differences give clues to current competitive positions and also can highlight weaknesses and opportunities that can be exploited in future strategy.

Also look closely at the activities and think about opportunities to increase the value added by creating performance measures for them. Measurements can lead the way to improvements. It is true what the business aphorism says: "What gets measured, gets done." Tool 86B lists sample performance measures in the value chain format.

Use value chain analysis to:

- Analyze business units based on the way they perform key functions.

- Look for competitive advantages through comparisons of the operations of competitors.

- Provide a formal model and presentation format for competitor research, analysis, and strategies.

As a recap, here are the five procedures to follow for value chain analysis.

1. Select the unit of analysis, both for your company and for major competitors.

2. Identify primary value-adding activities (direct, indirect, and quality assurance).

3. Identify support activities (human resources management and development; technology research, development, and design; and infrastructure activities).

4. Identify linkages between value chain activities.

5. Study the value chains to identify sources of competitive advantage.

Tool 86A

Sample Value Chain

INFRASTRUCTURE ACTIVITIES

	Transactions Processing	Reporting & Controls	Planning	Treasury & Risk Management	MIS

SUPPORT ACTIVITIES

Technology Research, Development, and Design

Human Resource Management and Development

PRIMARY ACTIVITIES

Purchasing, Inventory Holding, Materials Handling	Production	Warehousing & Distribution	Sales & Marketing	Customer Service

MARGIN

Tool 86B

Sample Performance Measures

INFRASTRUCTURE ACTIVITIES

SUPPORT ACTIVITIES

Transactions Processing	Financial Management & Reporting	Business Plan. & Analysis (Production / Sales)	Treasury / Risk Management	MIS
Cost • transactions processing cost **Quality** • number of transactions in error **Time** • time to process transactions	**Cost** • net income/earnings per share • ROI, ROE, and ROA • cashflow from operations • changes in net working capital • contribution margin • fixed asset turnover **Quality** • closing accuracy • budget accuracy **Time** • no. of deadlines met (e.g., SEC filings, 10K, etc.) • no. of man-hours to close	**Cost** • capacity (machine) utilization **Quality** • dollar var. from sales forecast • unit var. from sales forecast • margin var. from sales forecast **Time** (see Production and Customer Service)	**Cost** • cost of accounts receivable • current ratio • capital expense utilization • interest expense/coverage • EBIT • foreign exchange exposure • risk exposure • stock price • EVA **Quality** • bad debt expense • accounts receivable write-downs **Time** • A/R and A/P turnover	**Cost** • MIS cost • MIS personnel training cost • MIS return on investment **Quality** • cost saving from productivity improvements **Time** • project tracking

Technology Research, Development, and Design	**Cost** • R&D total cost • number of R&D projects entered into production	**Quality** • return on research and development		**Time** • man-hours per R&D project • payback period of R&R projects

Human Resource Management and Development	**Cost** • training expense • wage and benefits expense • workers compensation expense	**Quality** • number of accidents • time lost to accidents • training effectiveness • associate satisfaction		**Time** • training time • turnover • absenteeism

PRIMARY ACTIVITIES

Purchasing, Inventory Holding, Materials Handling	Production	Warehousing & Distribution	Sales / Marketing	Customer Service
Cost • total actual cost/vol. of purchases • raw materials inventory aging • materials handling labor cost **Quality** • raw materials rejects • number of on-time deliveries • raw materials spoilage, waste, etc. **Time** • number of late deliveries (from suppliers) • number of on-time deliveries (from suppliers)	**Cost** • total manufacturing cost • fin. goods/WIP inventory cost • no. of fin. goods inventory returns • no. of items in finished goods **Quality** • manufacturing performance • yield/volume produced (quantity) • waste produced • damaged WIP inventory **Time** • average cycle time • unscheduled downtime	**Cost** • distribution cost • transportation cost **Quality** • order accuracy • inventory fill rate **Time** • on-time delivery • warehousing order cycle time	**Cost/Revenue** • bookings margins • bookings orders quantity (units) • bookings cost • sales contribution margin • sales margin • sales revenue • sales cost • sales credits/allowance • returns (units, $) • canceled orders • selling, gen. and admin. cost • backlog in $, units **Quality** (see Production) **Time** (see Customer Service)	**Cost** • cost of inquiries **Quality** • customer satisfaction • number of customer inquiries • number of correct orders **Time** • inquiry resolution time • on-time delivery

MARGIN

179

13. Finance

This chapter contains case studies of how to present more detailed financial concepts.

Tool 87. Capital Investment Chart

By way of illustrating my next point, imagine being in an alternative universe in which electric power tools have not been invented. Builders put up houses using only handsaws and manual augers. Houses look pretty much the same as they do here, but there they take a lot longer to put up and they are much more expensive.

Then, one day while driving to your job at the hand tools manufacturing company, you have a brainstorm. You have the idea for power tools. It all comes to you in one complete vision, and you can imagine a power tools industry in all its glory: electric drills, electric saws, electric routers, and electric sanders. You are confident that as soon as the first power tools come on the scene, the demand for them will be explosive.

You are excited to get to work and to sell your idea to upper management. Since you are employed by a major industrial company, in order to get your idea approved, you must follow the procedures for capital investments called the capital investment guidelines.

The first step is to confirm the commercial viability of the project and provide a business commitment to start development. Determining the size of the potential market is the first order of business. In our example, what are the important factors in determining the size of the power tools market? Interest rates, construction activity, housing starts, age of existing homes, and demographics (growth in 35- to 54-year-olds) all might be factors to consider.

Key questions to be answered are:

- Is there a market need for this idea?
- Is the market big enough?
- Does the idea have global potential?
- Are the potential margins attractive enough?
- How much capital spending is needed?
- How much engineering time will be consumed?

- Who are the potential competitors, if any, and how big might they become?

- Does the idea make economic sense?

In order to get approval for this idea, the major industrial company that you work for requires that you write a project authorization request (PAR), which is a formal request for authorization. One of the schedules that this request must include is a quantitative financial analysis. Detailed calculations must be performed in order to calculate sales; cost of sales; selling, general, and administrative (SG&A) expenses; depreciation; taxes; investments; working capital; cash flows; internal rate of return (IRR); net present value (NPV); payback period; and sensitivity analysis.

These calculations can be seen in the capital investment chart (see Tool 87), which is broken down into three sections.

1. Profit and loss

2. Cash flow

3. Financial returns

Since this project illustration has an estimated life of five years, all calculations are based on five periods, with the sixth year used for additional investment and/or working-capital requirements.

This financial summary illustrates that the power tools project will require an initial investment of $2.044 billion. It has an IRR of 41.4 percent and a payback period of 2.7 years. However, using the various scenarios for sensitivity analysis, such as −10 percent of sales price, +10 percent of cost, +10 percent of investment, −15 percent of unit volume, +10 percent of days sales outstanding (DSO), and −15 percent of inventory turns, the section on sensitivity analysis calculates the financial return for the IRR, NPV, payback, and return on sales. These calculations illustrate the risk of the project based on the scenarios. The ideal project will have a higher upside potential than downside risk. Projects that have the reverse should be looked at very carefully to see what other positive elements they offer. Reduction in sales and cost overspending creates the greatest risk to the financial success of the power tools project.

At the power tools company, proposals for capital projects are communicated to management via a PAR. This request must include focused statements reflecting the strategic direction supported by a wide assortment of research results, reports, and schedules justifying the project. Here is an example of how the executive summary portion of your request might read.

Tool 87

Capital Investment Chart

POWER TOOLS PROJECT FINANCIAL SUMMARY
(in Millions of US$)

PROFIT & LOSS	YEAR						
	0	1	2	3	4	5	6
SALES	0	15,161	15,161	15,161	15,161	15,161	0
COST OF SALES	0	8,536	8,536	8,517	8,517	8,517	0
GROSS MARGIN	0	6,625	6,625	6,644	6,644	6,644	0
GROSS MARGIN %	0.0%	43.7%	43.7%	43.8%	43.8%	43.8%	0.0%
SG&A EXPENSES	0	4,291	4,291	4,291	4,291	4,291	0
SG&A %	0.0%	28.3%	28.3%	28.3%	28.3%	28.3%	0.0%
OPERATING INCOME	0	2,334	2,334	2,353	2,353	2,353	0
OPERATING INCOME % OF SALES	0.0%	15.4%	15.4%	15.5%	15.5%	15.5%	0.0%

CASH FLOW	YEAR						
	0	1	2	3	4	5	6
TAXABLE INCOME	0	2,334	2,334	2,353	2,353	2,353	0
TAXES at 35% OF INCOME	0	(817)	(817)	(824)	(824)	(824)	0
TAXES ALLOCATION		693	693	824	824	824	
TAXES PAYMENT		(693)	(693)	(824)	(824)	(824)	
INVESTMENT	(2,044)	0	0	0	0	0	
WORKING CAPITAL	0	(2,948)	0	0	0	0	2,948
BOOK DEPRECIATION		346	346	346	346	346	314
TOTAL CASH FLOW	(2,044)	(1,085)	1,863	1,875	1,875	1,875	3,262
CUMULATIVE CASH FLOW	(2,044)	(3,129)	(1,266)	610	2,485	4,361	7,623

	SENSITIVITY ANALYSIS						
FINANCIAL RETURNS	Base Case	Sales −10%	Cost +10%	Investment +10%	Unit Volume −15%	DSO +10%	Inv Turns −15%
IRR	41.4%	21.2%	23.6%	38.8%	36.8%	38.4%	38.5%
NPV (DISCOUNT RATE = 10.5%)	4,082	1,650	2,085	3,961	3,261	3,987	3,994
PAYBACK	2.7	4.2	3.9	2.8	2.9	2.9	2.9
RETURN ON SALES	15.5%	9.3%	9.9%	15.3%	15.0%	15.5%	15.5%

TO: Senior Management

SUBJECT: Worldwide Power Tools

The attached PAR recommends a total of $2.044 billion in tooling and equipment in order to establish a Worldwide Power Tools business.

Worldwide Power Tools will compete in four strategic business units: Professional Power Tools, Consumer Power Tools, Outdoor Power Tools, and Accessories. The combined served worldwide market size will be over $15 billion.

Professional Power Tools will be dominated by three or four global competitors. We expect that the Consumer Power Tools business will be concentrated in North America and Europe and that we will be the leader in the marketplace.

Outdoor Power Tools and Accessories will complement our other product offerings and be distributed through our existing channels.

This is a vitally important project for our existing hand tools products group, and timing is important. We urge your concurrence to this PAR and approval by the Board of Directors at the next Board meeting.

As the evening gets late and you pore over the numbers you have penciled in on the financial summary sheet before you, you imagine a kind of electronic calculating machine that could add columns of numbers and—okay, now back to this universe. The fact of the matter is that during our careers, few of us will ever identify a market opportunity as vast as the entire power tools industry (although personal computers have been invented in our lifetimes, and established companies such as Digital Equipment Corp. passed on the opportunity at a key moment). While we may never have the opportunity to prepare such a monumental business analysis, many of us will have the chance to do smaller projects. Working through this example of justifying an investment in power tools has revealed many lessons in how to prepare a business case for any capital investment, large or small.

Tool 88. Stair Chart

A stair chart compares a current five-year plan with previous five-year plans. This is accomplished by vertically aligning each of the plan estimates for a given year. (For example, all plan estimates for 2002 are aligned.) Each chart generally compares only one aspect of the plan, such as projected sales, projected capital expenditures, or projected contributions. By comparing the current projection for the coming year(s) with those made in previous years, one sometimes can determine such things as whether the plans have become more or less optimistic and whether the forecasts are consistent or erratic. By comparing the projections of multiple years, sometimes users can recognize basic changes in strategic thinking. Tool 88A depicts in general terms the overall layout of the stair chart.

Tool 88B is a stair chart for a corporation that has a five-year strategic planning horizon. During 1997, it developed a five-year strategic plan that contained sales forecasts for the upcoming five-year period (1998 through 2002). As can be seen by reading across the "1997 plan" row of the chart, it projected sales for 1998 to be $517.0 million, sales for 1999 to be $573.5 million, and so on.

A year later, during 1998, the corporation again produced a five-year strategic plan, which again forecast sales for the upcoming five-year period (now 1999 through 2003). The 1998 strategic plan was more optimistic than the 1997 plan: As you can see by comparing the "1998 plan" row with the "1997 plan" row, the sales forecasts across the board were now higher for each year than they were. For example, the sales estimate for 1999 had increased to $602.2 million from $573.5 million.

Tool 88A

Stair Chart

3 years ago 1998	2 years ago 1999	Last year 2000	Current year 2001	Coming year 2002	2 years out 2003	3 years out 2004	4 years out 2005	5 years out 2006
4 years ago ('97) plan for 1998 sales	4 years ago ('97) plan for 1999 sales	4 years ago ('97) plan for 2000 sales	4 years ago ('97) plan for 2001 sales	4 years ago ('97) plan for 2002 sales				
	3 years ago ('98) plan for 1999 sales	3 years ago ('98) plan for 2000 sales	3 years ago ('98) plan for 2001 sales	3 years ago ('98) plan for 2002 sales	3 years ago ('98) plan for 2003 sales			
		2 years ago ('99) plan for 2000 sales	2 years ago ('99) plan for 2001 sales	2 years ago ('99) plan for 2002 sales	2 years ago ('99) plan for 2003 sales	2 years ago ('99) plan for 2004 sales		
			Last year ('00) plan for 2001 sales	Last year ('00) plan for 2002 sales	Last year ('00) plan for 2003 sales	Last year ('00) plan for 2004 sales	Last year ('00) plan for 2005 sales	
				Current year ('01) plan for 2002 sales	Current year ('01) plan for 2003 sales	Current year ('01) plan for 2004 sales	Current year ('01) plan for 2005 sales	Current year ('01) plan for 2006 sales

Tool 88B

Stair Chart—Specific

($ Millions)	1998 sales	1999 sales	2000 sales	2001 sales	2002 sales	2003 sales	2004 sales	2005 sales	2006 sales
1997 plan:	517.0	573.5	674.6	759.1	788.8				
1998 plan:	542.9	602.2	708.3	797.1	828.2	836.5			
1999 plan:	542.9	584.1	687.1	773.1	803.4	811.4	820.0		
2000 plan:	542.9	584.1	721.4	811.8	843.6	852.0	861.0	950.0	
2001 plan:	542.9	584.1	721.4	852.4	885.7	894.6	904.1	997.5	1,000.0

During 1999, the annual strategic planning cycle was repeated once again. This time the five-year plan was less optimistic than it had been the previous year. The sales estimates for each year were lower than they had been the previous year. For example, the year 2001 is now forecast to come in at $773.1 million vs. a prior estimate for the year of $797.1 million.

Tool 89. Performance Pyramid

In most companies, finance professionals have been trained to interpret and integrate large amounts of data. Unless it is disseminated clearly, data may be misinterpreted by persons with limited understanding of how it relates to a company's objectives. The performance pyramid (see

Tool 89) is one way to avoid misinterpretation by demonstrating that the various components of the income statement and the balance sheet are building blocks that fit together in one common structure. Sales, gross margin, and SG&A expenses are included among the building blocks that make up the income statement side of the pyramid. Working capital, fixed capital, and debt are the building blocks on the balance sheet side of the pyramid.

A manager who has a specific objective involving one of these blocks gains a better understanding of its contribution to the overall financial plan within the context of the pyramid. For example, a vice president (VP) of sales who has a sales objective or a VP of manufacturing who has a productivity or gross margin objective can use the pyramid to determine where those objectives fit into the company's key objectives. Sales data is transformed into information on company revenues, and production data becomes part of total costs and expenses.

At the top of the pyramid is the company's financial performance target: in this case, return on net assets (RONA). Depending on the audience for the presentation, this measurement just as easily could have been return on equity or economic value added. Every item below the apex of the pyramid is a value driver that helps shape the top measurement and determines whether that measurement is improving or not and its rate of change.

Measurements overemphasizing one area at the expense of another can lead a company off course. Cash flow straddles the income statement and the balance sheet because its measurement is driven by the effective management of these important financial tools, and it is the foundation on which a business rests. The pyramid is a one-page summary that nonfinancial managers can interpret easily. By focusing on measurements that straddle the income statement and balance sheet, financial managers help create a balanced and integrated approach toward planning and budgeting. Using the performance pyramid helps close the loop on financial measurement by ensuring that no important elements are overlooked. As a presentation format, the image of the pyramid is powerful, enduring, and simple—often more compelling than raw facts and figures.

Tool 90. Sources and Uses of Funds Chart

In the February 1995 edition of *Management Accounting* magazine, I wrote an article entitled "The One-Page CFO." In that article, I argued that even the most complex financial issues can be boiled down to one-page presentations. I provided several examples of this, showing how to simplify

Tool 89

Performance Pyramid

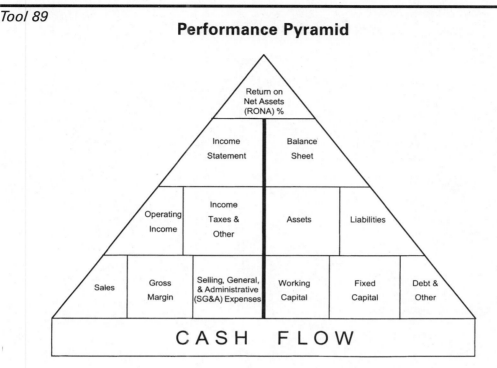

presentations of strategic plans, grants of authority, and capital expenditures. These examples were geared to the large corporation.

During the summer of 1995, I was approached by the Regional Economic Studies Program of the University of Baltimore to simplify a financial presentation in the public sector. The Regional Economic Studies Program had been asked by the Maryland Department of Human Resources (DHR) to help them gain a better understanding of their finances. While analyzing the following case study, I found the challenges inherent in presenting public sector issues on one page to be equal to any I had seen in the private sector.

The DHR was concerned about upcoming welfare reform proposals, including proposals to replace the federal guarantee for Aid to Families with Dependent Children (AFDC) benefits with a work requirement and a time limit; to allow states the option to block grant child welfare programs; to cut food stamp funding for nonimmigrant programs; to block grant administrative costs for school lunch programs; and to thwart various provisions concerning immigrants. In order to prepare for the upcoming changes, DHR wanted to better understand the current sources and uses of funds for its various programs. The end result of my assignment was to develop baseline descriptions of four programs, with

Tool 90

Sources and Uses of Funds Chart

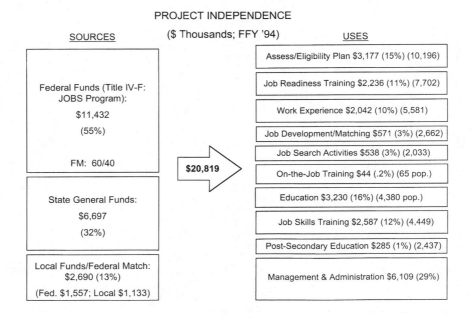

PROJECT INDEPENDENCE

($ Thousands; FFY '94)

SOURCES / USES

SOURCES

Federal Funds (Title IV-F: JOBS Program):
$11,432
(55%)

FM: 60/40

State General Funds:
$6,697
(32%)

Local Funds/Federal Match:
$2,690 (13%)
(Fed. $1,557; Local $1,133)

$20,819

USES

Assess/Eligibility Plan $3,177 (15%) (10,196)

Job Readiness Training $2,236 (11%) (7,702)

Work Experience $2,042 (10%) (5,581)

Job Development/Matching $571 (3%) (2,662)

Job Search Activities $538 (3%) (2,033)

On-the-Job Training $44 (.2%) (65 pop.)

Education $3,230 (16%) (4,380 pop.)

Job Skills Training $2,587 (12%) (4,449)

Post-Secondary Education $285 (1%) (2,437)

Management & Administration $6,109 (29%)

an accompanying one-page graphic showing the sources and uses of funds for each program.

Project Independence (PI) is Maryland's version of the federally mandated JOBS program, which is a federal- and state-funded mandatory employment training program for AFDC recipients to assist them in attaining unsubsidized employment. The services available to PI participants include job skills training to prepare for special occupations; job readiness education to teach clients good work habits; and job search assistance for clients seeking job skills and placement. The program is required to serve all AFDC clients with children over age three.

There are three sources of funds for PI. The primary funding source is through Title IV-F (the JOBS program), consisting of 60 percent federal funds and 40 percent state matching funds. All funds from these sources finance 10 separate programs. Tool 90 shows the sources and uses of funds for PI.

Tool 91. Hockey Stick Graph

Never aspire to manage to "hockey stick" plans; if the revenue projections resemble a hockey stick graph, then this is one indication of a business to avoid. The most common action taken by people in lousy businesses is to pretend that they are in a great business; it is also the most deadly. Most managers want to believe their business has growth potential even when it does not. Many have gone to great lengths to deny the realities of their markets. One particular multibillion-dollar, single-business company illustrates exactly how much managers want to believe that they are in a growth business. When this company plotted its revenue projections from its last five consecutive five-year plans, the results looked like a hockey stick graph. (See Tool 91.)

When the results were shown to the top 17 managers in a strategy session, several almost fell off their chairs. They could not believe they had not been in a growth business all these years. They had never taken the time to look at their historical projections.

When the company's actual revenue performance was laid on the "hockey stick" projections, guess where they fell? Of course, they fell along the flat portion of each hockey stick.

The management team had believed each year for five years that their business was in a short-term slump and that revenue growth would resume soon. They had a business strategy in place that would position them for the expected upturn when it occurred. As a result, they were carrying costs for excess capacity in people, plant, and equipment that were destroying their bottom line.

Once they realized that they were in a mature business, they made major changes in their business strategy to make it consistent with their business's growth potential. They cut excess capacity and people to size their business to match its market. They quickly improved service to retain their current customers. Profit margins improved.

Businesses eventually decline. Most managers do not like to be in declining businesses. A business that is getting smaller has less opportunity—unless, of course, you want to take advantage of the opportunity to learn how to lay off people and cut other expenses in order to preserve profitability. In one company that was being "rightsized," the standard joke among middle managers was that the new incentive system defined a "top performer" as someone who eliminated his or her own job.

While pretending to be in a great business may seem like a crazy idea, more than one manager has done so to make it through his or her remaining time to retirement. The crumbling walls were left for someone else to repair.

"Hockey stick" plans abound in businesses throughout the world. Managers regularly use them to preserve their careers. While they are usually not in the best interests of the shareholders, such plans are part of the game played by managers worldwide.

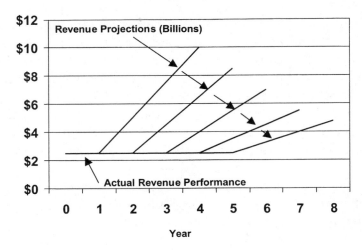

Hockey Stick Graph

Tool 92. Dashboard

Complexity is the bugaboo of business. Because firms must factor so many variables, most do not know how to monitor their performance optimally. Questions abound and there are no simple answers. Which metrics are most essential to monitor? How frequently should these metrics be sampled? What is an appropriate alarm threshold?

Traditionally, business monitoring tools have focused on alerting managers to problem situations after some alert threshold has been reached, such as "disk space at 85 percent." Too many managers spend too much time putting out fires instead of preventing them. This process is mostly reactive, forcing managers to wait until the monitored data reaches some near-critical state before responding. In simple terms, managers need to be proactive, not reactive.

The dashboard is an attempt to put all of the relevant measurements for a corporation on one sheet of paper. (See Tool 92A.) The idea is that just as a driver of a car needs certain information to drive a vehicle, so, too, does the corporate manager need certain information to steer a corporation. This analogy can be extended one step further. In the automobile, certain metrics need to be monitored continuously and reported on continuously, such as the speed the car is traveling and the amount of fuel in the tank. Certain metrics need to be monitored continuously but reported on only periodically (whenever there is a problem): for example, when the wiper fluid is low, or when a seat belt is not fastened. Other elements need to be sampled and reported on periodically, such as the alarm that sounds when the lights are left on after the engine is turned off. So, too, on the corporate dashboard, the combination of gauges, meters, and lights can be customized to meet the particular needs of the entity being measured. When establishing the dashboard, it is important to consider the reporting frequency and the thresholds for all of the measurements.

Whatever the final format, it is a good idea to clearly spell out the definitions for the various meters and gauges included on the dashboard.

Tool 92B is the definition layer behind this corporate dashboard. In essence, it reveals the wiring behind the dashboard.

Tool 92A

Dashboard

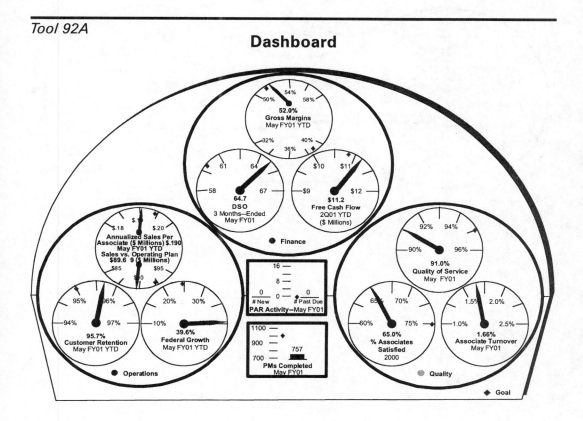

Tool 92B

Dashboard Definition Layer

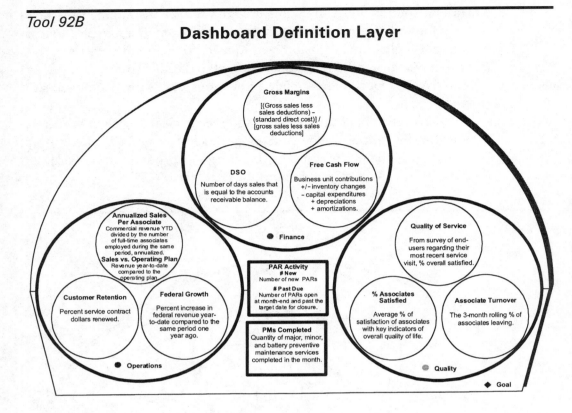

14. Marketing

This chapter contains case studies of how to present sales and marketing analysis and marketing positioning.

Tool 93. Marketing Strategy Matrix

A firm's context, or position in its market, has significant implications for strategy, and the marketing strategy matrix is a prescriptive tool that links marketing strategy to market context. The matrix can be used to identify these implications and make sure the strategy reflects them. The marketing manager can use this tool for generating strategy or for reviewing strategies proposed by staff or operational managers. The chief executive may find this tool useful in checking the internal consistency of marketing plans and proposals, as it can be used to compare the context and the strategy as defined in a marketing plan. The prescriptions of the matrix are based on a compilation of many studies of marketing strategy, and while not 100 percent valid in every situation, they can be expected to be valid in most circumstances. Deviations from the prescriptions of the matrix should be justified by an explanation of how the situation differs from the norm.

In Tool 93, we have created a marketing strategy matrix that links five possible strategic situations with six strategy options, for a 5 by 6 matrix. Here are the instructions to follow in order to create the marketing strategy matrix.

    A. Identify the context by delineating possible strategic situations. We have chosen to illustrate five strategic situations on our matrix.

        1. *Market development.* Early entry or technical leadership makes a company a pioneer in its market.

        2. *Market domination.* The leader in an established market. Holds an advantageous and influential position.

        3. *Differentiated advantage.* A firm (not necessarily the leader) has a sustainable advantage such as low cost or patent protection.

        4. *Market selectivity.* Characterized by segmented buyer wants and many small firms addressing these wants. Local service

Tool 93

Marketing Strategy Matrix

STRATEGIC SITUATIONS

STRATEGY OPTIONS	Market development	Market domination	Differentiated advantage	Market selectivity	No advantage
New-product development	Yes	Yes. Develop product portfolio	Consider	Yes	No
Segmentation and targeted marketing	Yes. Top priority markets first	Yes. Target many segments	Yes. Make sure focus is right	Yes. Make sure focus is right	Yes. Narrow focus
Product positioning through research and marketing	Establish marketing mix first	Cover multiple positions	Position to highlight advantage	Only if current position is ineffective	May not be cost-effective
Sales and marketing productivity improvement	Delay	Yes	Yes	Yes	Consider
Acquisition or merger	Questionable	Consider	Consider	No. Evaluate acquisition threats	Yes
Harvest or divest	No	No	Questionable	Consider	Yes

businesses such as restaurants are good examples (segmented based on price, menu, and location).

5. *No advantage.* The other situations all imply a strategic advantage or at least the potential to create one. Sometimes there is no immediate or obvious basis for creating an advantage.

B. List generic strategy options in the matrix. We have chosen the following six strategy options.

1. New product development

2. Segmentation and targeted marketing

3. Product positioning through research and marketing

4. Sales and marketing productivity improvement

5. Acquisition or merger

6. Harvest or divest

C. Fill in the matrix at the intersections of each strategic situation and strategy option.

Use the marketing strategy matrix to:

• Identify feasible strategic options given a specific market context

- Evaluate marketing plans and proposals to see if their underlying strategies are realistic given the marketing situation

As a recap, here are the procedures to follow when using the marketing strategy matrix.

- Select one of the five generic strategic situations that best describes the marketing context for your business and product.

- Use the matrix to identify applicable strategy options or to see whether a strategy is appropriate for the given situation.

Tool 94. Product/Market Planning Matrix

Amana uses a product/market planning matrix when deciding what types of new products to develop. The roots of Amana's matrix are found in Ansoff's product/market expansion grid. (See Tool 94A.) Ansoff's grid identifies four strategies.

1. Market penetration

2. Market development

3. Product development

4. Diversification

Amana has added more cells to the matrix (see Tool 94B) because it found that product development focuses on improving old products and developing related products. Diversification into unrelated products and markets is risky. Amana has integrated the matrix into its product planning to ensure that all possible opportunities are considered.

To use the matrix, first lay out a table similar to the one in the product/market planning matrix and fill in your current products and markets. In Tool 94B, Amana lists microwave ovens to appliance stores in the first cell, "Present Products/Same Markets," since it currently sells microwave ovens through appliance stores.

The next and more difficult step is to fill in the rest of the cells. Amana tries to specify opportunities for each category even though every opportunity is not necessarily pursued. Usually a variety of existing proposals can be entered. If these do not fill all the cells of the matrix, ask the product development and marketing people to go back to the drawing board. A brainstorming session is a good way to identify opportunities. The marketing department should provide a breakdown by segments, such as type of buyer, geographic area, or distribution channel.

Tool 94A

Ansoff's Product/Market Expansion Grid

	Current Products	New Products
Current Markets	1 Market penetration strategy	3 Product development strategy
New Markets	2 Market development strategy	4 Diversification strategy

Tool 94B

Product/Market Planning Matrix

	Present Products	Improved Products	Related New Products	Unrelated New Products
Same Markets	Microwaves to appliance stores	Less expensive microwaves	Combination microwave and electric oven	?
New Markets	Microwaves to department stores	Built-in microwaves	Commercial microwave ovens	?

If you have a long product line, you cannot fit all the current products and markets on a single matrix. Use the matrix as a tool for presenting the concept to your staff and associates, but switch to lists to record the information. Create a list for each cell of the matrix, or do a separate matrix for each family of products or product lines.

The final step is to select some of the opportunities for development and introduction. Most companies have a standard procedure for selecting projects but no standard procedure for identifying opportunities. Use the matrix to identify opportunities that might not have been considered otherwise; then feed all those opportunities into the standard evaluation process. Typically, this means product development or marketing managers select the most promising for market research and technical review, followed by formal proposals to senior management.

Use the product/market planning matrix to:

- Identify product development opportunities.
- Systematize the search for new product ideas.
- Present new product opportunities visually.

As a recap, here are the four procedures to follow when using the product/market planning matrix.

1. Draw the matrix. If you have many products, use separate matrixes or lists for each product family.

2. Enter current products into the "Present Products/Same Markets" cell of the matrix.

3. Fill the rest of the cells with new product ideas. (A brainstorming session is helpful.)

4. Evaluate the opportunities identified and pursue those that have merit.

Tool 95. Product/Market Certainty Matrix

The product/market certainty matrix is an outgrowth of Ansoff's product/market expansion grid. However, this method adds the concept of risk to come up with an innovative application for sales forecasting.

The probability of achieving a sales forecast for a product will tend to be lower if the product is new or if it is to be sold to a new market. However, most sales forecasts are not adjusted for risk, and as a result resource allocation and financial projections can be way off for high-risk products. Use this method for risk-adjusted forecasts of product sales for a one-year or five-year plan. It is especially useful for companies with frequent product introductions.

Here are the instructions to follow to use the product/market certainty matrix. Start by categorizing each product (or service) by the following two criteria.

1. *Market.* Existing; new but related to existing; or new/unrelated

2. *Product.* Existing; new but related to existing; or new/unrelated

Draw a 3 by 3 matrix (see Tool 95) with product categories on the horizontal axis and market categories on the vertical axis. Enter all the products in the appropriate cells of the matrix (or on separate lists if there are too many products to put on the matrix). Write down the sales projections for each product and sum all projections within each cell to obtain totals for each cell. Write these nine totals in the cells of the matrix.

Estimate the probability that these projections will be correct. If you have historical data for many products, use it to find out what percent of the time forecasts was close to performance in previous periods. If you break out the historical data into the same nine groups, you will be able to calculate the percentage of accurate forecasts your firm produced for each of the categories. However, many companies will find that they do not have a large enough sample of product histories or that they have not saved the forecasts. (Incorrect forecasts have a way of disappearing.)

Alternative: Use conventional percentages for the cells. These range from a 90 percent accuracy rate for the existing product/existing market cell to 10 percent for the unrelated product/unrelated market cell. The conventional percentages appear in Tool 95.

Now that the matrix shows sales forecasts and probabilities for each cell, you can use it to see how risky those sales forecasts really are. To do this, calculate a total forecast—the sum of all the product forecasts. Then calculate a risk-adjusted total by multiplying each forecast in the matrix by the percent figure appearing in its cell before totaling the matrix forecasts. If all the projections are for existing products and markets, then the risk-adjusted total will be only 10 percent lower than the first total. But if higher-risk products are included, then the risk-adjusted total may be considerably lower. Use this figure as a pessimistic forecast to model the impact of lower sales on financial performance. If the effect would be catastrophic, you might consider a more conservative financial strategy.

You also can use the matrix to identify high-risk projections and plan follow-up as needed. You probably can reduce the risk of a five-year projection for an unrelated product in an unrelated market by reassessing it a few months after introduction. Conversely, more information may increase the risk factor from 10 to 20 percent or more—this is a judgment call for management. (*Key point:* It does not make sense to live by a forecast for five years, or even one year, if it has only a 10 percent chance of being correct! It is better to treat these forecasts as temporary and permit them to be modified whenever new information is available.)

Use the product/market certainty matrix to:

- Adjust long-term sales forecasts to reflect the current risk associated with each product or service.

Tool 95

Product/Market Certainty Matrix

MARKET \ PRODUCT	Existing	New/Related	New/Unrelated
Existing	90% 450,000 (X 90% = 405,000)	60% 200,000 (X 60% = 120,000)	30% 0
New/Related	60% 100,000 (X 60% = 60,000)	40% 175,000 (X 40% = 70,000)	20% 50,000 (X 20% = 10,000)
New/Unrelated	30% 150,000 (X 30% = 45,000)	20% 0	10% 0

Total Forecast = $1,125,000

Risk-Adjusted Forecast = $710,000

Overall Risk Factor = $710/$1,125 = 63%

- Analyze revenue projections to identify their sensitivity to risk.
- Identify product sales forecasts that are least likely to be accurate.

As a recap, here is the four-step procedure to follow when using the product/market certainty matrix.

1. Categorize each product by market and product criteria.
2. Prepare a 3 by 3 matrix labeled with product and market categories. List the appropriate products for each cell.
3. Sum the product forecasts for each product in a cell and enter this total in the cell. Fill in all of the cells of the matrix with sales-forecast figures in this manner.
4. Adjust these figures for risk, referring to your firm's historical experience or using conventional risk percentages.

Tool 96. Market Map

Most market research surveys collect information on consumer awareness of a product, consumer usage of the product, and similar statistics on the competition. Studies often break consumer behavior into unaided and aided awareness, trial, usage, and preference, and sometimes into even more detailed categories.

This information is useful in developing marketing strategies and tactics. For example, if a product has a low awareness figure (say only 15 percent of the market knows it exists), then the strategy should be to build awareness, probably through educational advertising and promotion. But when a company performs multiple surveys on many products, often using multiple research firms, a large, confusing collection of awareness and usage statistics is created. Management can find it difficult to sift through the reports and compare statistics to evaluate the appropriateness of a proposed strategy or tactic. The market map provides a simple, visual summary of these statistics, making it easier for management to utilize survey data for strategy development and oversight.

Here are the three steps to follow when using the market map.

1. You need good data on awareness, trial, and usage of the brands that you want to map. In general, marketing departments of larger companies already collect this information at least annually, but some industrial companies and smaller firms will not have current statistics. In this case, seriously consider hiring a research firm to perform a basic benchmark study. As long as you tell it to confine the research to the statistics needed, the study should not prove too expensive. A survey should include the following.

   • Unaided awareness (asks consumers to name brands/products/companies in the market)

   • Aided awareness (asks if they have heard of specific brands)

   • Trial (asks if they have tried the brand recently)

   • Usage (asks if they use the product regularly)

   • Preference (asks if they prefer the product over competitors)

   Sometimes studies break consumer behavior down into many more categories, depending on the model of consumer behavior and product adoption that the company and its research firm decide is most appropriate.

   You must study a random sample of the potential users of your product, of course, in order to produce legitimate results. A telephone survey will give better results than a mail survey in most cases because the response rate will be much higher; thus the chances of respondents providing a fair representation of the original sample is higher. In addition, the name of your company and product should not be associated with the survey—this is the main reason why independent research firms usually perform these surveys.

2. Create a market map for your product. The market map simplifies survey results slightly by breaking consumers into the following categories.

- Those who are aware of the product

- Those who have tried the product and prefer it

- Those who have tried the product but have no strong preference—they are indifferent

- Those who have tried the product and do not currently use it—they have in essence rejected it

Review your data to identify the percentage of consumers falling into each of these categories.

Note: These categories may or may not correspond to the data you have to work with. But simple logic should be sufficient to translate any study into these categories. For example, the consumers who report trial of a product but do not report current, regular use or preference for it make up the group of rejecters—even if the survey did not specifically ask consumers whether they "rejected the product."

Next, create a graph with both axes ranging from 0 to 100 percent. Label the horizontal axis "Awareness" and draw a vertical line from it to represent the percentage of consumers who are aware of the product.

Label the vertical axis "Trial" and draw a horizontal line representing the percentage of consumers who tried the product (including those who have rejected it, are indifferent to it, and prefer it). Now divide the resulting box with additional horizontal lines to break down the group of triers into their three constituent groups. (Use the scale of the vertical axis to keep these areas in proportion to the rest of the graph.)

Tools 96A and 96B show two steps in the creation of a market map—the first shows awareness and trial, and the second breaks down trial into its components.

3. Interpret the map. Use the map to identify and illustrate the key challenges and issues facing a product. (Prepare a time series of maps and maps for the competition in order to illustrate trends and competitive issues.)

The examples are for a product that has a 70 percent rate of awareness and a 65 percent trial rate. This suggests that the company has done a good job of converting awareness to trial and does not need to focus on this in future marketing. Building awareness has proven to be a good strategy, and marketing communications should adopt this as an objective.

When the triers on the map are broken down into those who prefer, are indifferent to, or have rejected the product, other objectives are indicated as well. Only 25 percent of all who are aware of the product in the survey were users, and only 5 percent preferred the product. These statistics point to a need to encourage usage among those who try the brand, perhaps through promotional incentives to increase the trial period. The statistics also may indicate a lack of product strength. Advertising should be used to emphasize product benefits, and the product should be compared with its competition to see whether the low preference rate reflects a product weakness or simply a competitive market. The market map is a great tool to identify areas that need further study.

Use the market map to:

- Translate awareness and usage data into a visual display as an aid to developing marketing strategies and plans.

- Compare competing products or a portfolio of products on the basis of consumer awareness and usage.

- Present survey research data to senior managers.

As a recap, here are the three procedures to follow for creating the market map.

1. Obtain data on consumer awareness and usage of the product(s) in question. (Perform market research if necessary.)

2. Create a market map for each product or brand by plotting trial percentages on the vertical axis of a graph and awareness on the horizontal axis.

3. Use the market map to develop marketing strategies and plans and to present survey statistics in a visual format.

Tool 96A

Market Map—Step One

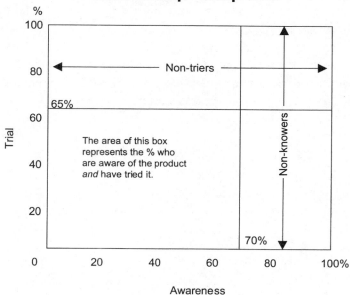

Tool 96B

Market Map—Step Two

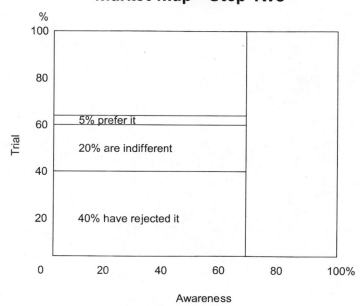

The only problem with the market map is that it contains so much information that it takes a little study to get used to the format. Once you are familiar with it, however, you should find it a simple and useful aid to marketing decisions.

Idea: Senior managers may want to request that the marketing department present all awareness/trial data from surveys in a standardized market map format to make interpretation easier and to compare rate of change over time, or between product lines.

15. Performance

This chapter contains case studies of how to present various other business performance measurements.

Tool 97. Sales Grid

The sales grid is an example of a business matrix where labels are shown in each quadrant to describe or categorize entities in that quadrant. In our example, Tool 97, the sales grid is used to evaluate salespersons.

Left to their own devices, most salespeople fall into a standard style of selling that is most comfortable for them. Quotas and other quantitative measures of performance do not provide direct feedback about the effectiveness of selling style, so salespeople and managers tend not to evaluate the appropriateness of style as often as they should.

The sales grid is a powerful tool for analyzing the performance of individual salespeople, training salespeople to identify and improve their selling styles, or planning long-term efforts for complex sales. The use of the grid in planning is rare; rather, the grid is most effective in assessing the current environment.

This grid uses concern for the customer on one dimension and concern for the sale on the other. These represent the two most important dimensions of sales style. The traditional hard sell is a totally sales-oriented style, with no customization of message or approach, with only concern for sale. This is stereotypically the "American" approach to sales. The other extreme is a personal style, in which long-term personal relationships are developed and are necessary prior to any sales (the so-called Far Eastern approach to sales), also referred to as concern for customer. The concept of consultative selling, involving a long-term, problem-solving relationship with a client company, is represented on the grid as a combination of high concern for both customer and sale.

To plot a particular selling style on this grid, first rate the salesperson's concern for the customer on a 1-to-9 scale (1 equals low, 9 equals high). It may be helpful to rate a group of salespeople at the same time in order to add a comparative element to the rating. Next, rate the salesperson's concern for the sale on a 1-to-9 scale.

Note: It is preferable to *observe* salespeople in action before rating selling styles.

Use the two ratings to plot each salesperson's style on the grid and identify the generic strategy that best describes him or her. The strategies are indicated in Tool 97, and their definitions are as follows.

- *People-oriented.* Builds a personal relationship with the customer. Motivated by desire to be liked. Building rapport helps sway the customer to purchase.

- *Problem-solving–oriented.* Studies customer's needs and sells products that satisfy these needs. Can involve building a consultative relationship with the customer in which the customer purchases because the salesperson can foresee or solve problems better than others.

- *Sales technique–oriented.* Combines approximately equal problem solving and personality-oriented appeals.

- *Take-it-or-leave-it–oriented.* Allows the product to "sell itself" without making an effort to get to know the customer or to push the product to the customer. This is a passive selling style.

- *Push-the-product–oriented.* Tries to corral the customer and create pressure to purchase through hard-sell tactics. Emphasizes product features rather than personal relationship or customer needs. This is the most aggressive style. (May include reference to customer needs and problem solving, but in a generic and not always credible manner.)

Identifying the current styles of the people in your sales force may provide useful clues as to what makes some more effective than others. Companies often look to top performers for lessons in how to sell, and in many cases, top performers are successful because their style ranks high on one or both dimensions of the grid.

Salespeople should be encouraged to identify and develop the style that is most comfortable and effective for them and to try to stick to this personal goal in most selling situations. They also should recognize that they probably have a backup style they use in difficult and tense situations. This tool can be modified by targeting an ideal backup from the grid. Any combination of styles is possible when selecting ideal and backup styles, but clearly styles that are closer to 9,9 are preferable. In general, the 1,1 style is going to be ineffective.

As a recap, here are the five procedures to follow when using the sales grid.

1. Observe sales staff in action.

2. Evaluate each salesperson's concern for the customer on a 1-to-9 scale.

3. Evaluate each salesperson's concern for the sale on a 1-to-9 scale.

4. Plot each salesperson's style on the grid and identify which of the five generic styles describes it.

5. Use the grid as a training tool.

Options: Use the grid to teach selling styles or to evaluate style and performance of salespeople. Categorize customers and prospects according to the sales style that they prefer. It is also possible to map the life cycle of a complex sale on the grid and use the grid to develop a strategy for such sales.

Tool 97

Sales Grid

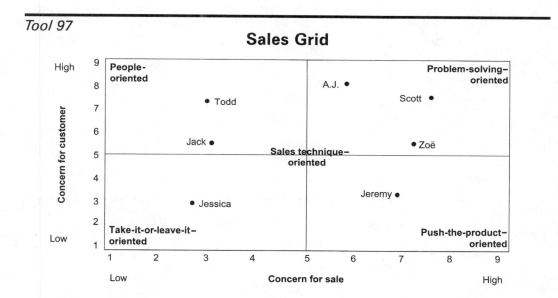

Tool 98. Cost/Performance Matrix

James River Corp., a paper manufacturer, faces a competitive industry in which many sales are commodity-oriented and price is typically very important. The tendency in industries like this is for competitors to focus on price and volume. Innovation often centers on cost reductions. But because scale has an impact on cost in paper manufacturing, many smaller mills have found it difficult to compete. Some, however, have differentiated their products on the basis of quality and have targeted segments with specialized needs.

James River has developed a product planning tool that helps ensure every new product offers a competitive combination of quality and price. This method focuses attention on performance but does not break performance down into the many individual attributes that most marketing

Tool 98A

Cost/Performance Matrix

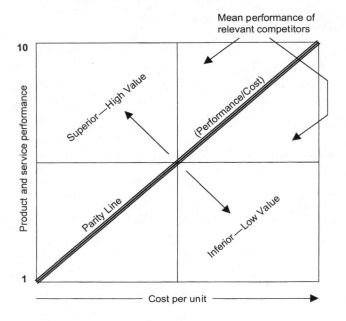

Mean performance of
relevant competitors

Tool 98B

Cost/Performance Matrix—Specific

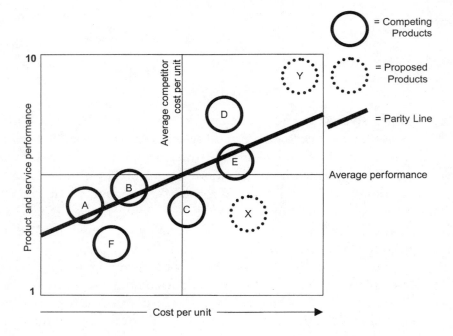

studies delineate. Instead, the method puts performance in the context of a price to create a value map. James River's cost/performance matrix defines performance from the customer's perspective—determined through formal surveys or informal interviews. The other dimension of the matrix, price, is represented by manufacturing cost, since cost is the best predictor of pricing strategy in the long run in price-sensitive markets. Research and development (R&D) and/or manufacturing often can provide good estimates of the relative cost position of competing manufacturers, and product characteristics provide additional information about the unit costs.

Here are the four steps to follow when using the cost/performance matrix.

1. Survey customers to obtain rankings of overall product and service performance on a 1-to-10 scale. (James River uses a consultant, which lends objectivity to the questions about competitors' products.) Ask customers to provide overall ratings of your product and its major competitors. If the product is in the planning stages, then a focus group should be used rather than a telephone or written survey.

   Idea: Marketing may already have done a formal research study that shows which product attributes customers value most and how they rate each competing product on each attribute. Use this data by weighting attributes by importance, summing weighted ratings to give a single performance measure, then scaling this rating to the 1-to-10 scale of the matrix.

   Although market research is helpful in evaluating customer perception, a formal study is not required. Quizzing customers, the sales force, and distributors can provide a quick back-of-the-envelope estimate of each product's performance scale. *Caution:* Technical staff members tend to define performance according to objective laboratory measurements, but these should not be used in the matrix unless there is clear evidence that customers use the same measures.

2. Estimate cost per unit for your own product proposal and for each competing product. James River collects as much information as possible about the manufacturing facilities of competitors to provide a basis for these estimates. In other industries, distribution channels, marketing programs, and other costs also might be relevant to cost analysis. Prices are often a good clue to product costs as well, and the sales force and distributors will have a

good idea of standard markups and any competitive pressures that might suppress prices. Public companies may release data on revenues and earnings by segment. Finally, where economies of scale are important and quantifiable, cost can be deducted from volume or share.

3. Plot each competing product on a graph with performance on the vertical axis and cost on the horizontal axis. (See Tools 98A and 98B.) Draw a vertical line at average competitor cost per unit and a horizontal line at average performance. The intersection of these two lines represents the average value offered by current products and provides a general target for new-product development.

4. Draw a diagonal line to represent the various combinations of cost and performance that are equal trade-offs from the customer's perspective. This is called the parity line.

Note: You may estimate the slope of this line by performing a linear regression on the paired cost/performance data for the products in this market or by drawing a line by eye. Another approach is to do a market research study in which trade-offs are presented to customers and the customers are asked to identify combinations of price and performance that offer equal value. *Warning:* With new-product introductions, the parity line can move over time.

Products below the parity line probably need improvement or cost reductions to bring them up to or above parity. Any new-product offering should be aimed *above* the parity line.

The sample matrix (see Tool 98B) illustrates a hypothetical paper market with six major products at present, plus two product proposals that R&D is considering for possible development and production. The proposed products are labeled X and Y. X will offer high cost and below average performance, and Y will offer very high cost and performance. Product X is closer to existing products in both cost and performance. R&D and marketing are uncertain that the market will accept Y because it is considerably beyond competing products in both performance and price. However, although X is closer to current products, it falls below the parity line. Customers will probably view it as offering no more value than product C, which has a lower cost position. Product Y lies well above the parity line and could be perceived as a good value despite its high cost.

Use the cost/performance matrix to:

• Identify cost and performance requirements for a drawing-board product.

- Compare competing products on the basis of their value to the customer.

- Evaluate new products by comparing them to market norms for performance and value.

- Develop a quality- or performance-based strategy in industries where innovation is traditionally cost-oriented.

As a recap, here is the four-step procedure to follow when using the cost/performance matrix.

1. Rate products in a market or market segment on the basis of overall quality of the product and the services associated with it. (Estimate quality for your new and proposed products as well.)

2. Estimate cost per unit for each of the products.

3. Plot the products on a cost/performance matrix. Add lines for average performance and cost, and fit a parity line to the data using linear regressions, or eyeballing.

4. Identify the products that fall significantly above the parity line (winners) and below it (losers). Map proposed products to see where they stand relative to current products.

Tool 99. Leadership Factor Matrix

Albrecht & Associates, a San Diego consulting firm, specializes in increasing the effectiveness of organizations through organization development. As part of the process, an organization assesses its performance in great detail, looking at everything from administration to innovation. One of the assessment tools Albrecht & Associates uses is a leadership factor matrix—a quick device for evaluating leadership style and effectiveness. Its simplicity and ease of use make it a good device for thinking about leadership issues or problems and doing appraisals, and managers will find it well suited to an informal, back-of-the-envelope analysis. (See Tool 99.)

Here are the four steps to follow when using the leadership factor matrix.

1. Collect information about the style of a manager, through observation, discussion with employees and associates, or a combination. For example, employees can be asked whether they think their manager gives them too much, too little, or the right amount of responsibility, and the results could be used to assess the effectiveness of the manager's delegation and to rank the manager on

the autonomy factor. Formal questionnaires can be used, but in many cases, observation or experience alone is sufficient to rank managers.

2. Decide which leadership factors, or standards, are important to management. Examples of factors that may be deemed desirable include being a goal sharer, a team player, an autonomy granter, or a reward giver.

3. Evaluate the observations and information from interviews to rank each manager's style for each desirable factor. This may involve sifting through a variety of complaints and observations to deduce where the manager ranks. For example, if you learn that the manager has not held a staff meeting for six months and that some employees resent seemingly arbitrary decisions made by this manager, then you would conclude that the manager placed a low emphasis on teamwork.

   Note: There is no absolute measure of the right emphasis—the rankings are relative to the priorities of the organization or the desires of the executives.

4. Next, use the matrix to map managers based on their leadership style. The matrix summarizes common behaviors for each factor, placing them on a spectrum from low emphasis to high emphasis. This gives a general orientation to the evaluator who wishes to make rankings conform to the norms established by Albrecht & Associates. It also gives some clues to what is considered an ideal emphasis for each factor—clues that may be of use in training or performance evaluations. Use the evaluation of the manager in placement, appraisal, training, and other leadership decisions. This process reduces subjective appraisals in favor of more standardized and understood company-wide priorities.

   Idea: If you are in a new position, use this matrix to chart the style of the manager you are replacing. Then compare it with your style. Significant differences will be sources of confusion for employees, and a gradual transition may be required on the factors where your style differs.

Use the leadership factor matrix to:

- Understand your own or someone else's leadership style.
- Diagnose leadership problems or complaints.

- Anticipate problems you may encounter in a new position due to differences between your style and the style of the previous manager.

- Evaluate leadership style for performance appraisals and management training.

As a recap, here are the four procedures to follow when using the leadership factor matrix.

1. Observe managers and/or interview their employees and associates.

2. Determine key factors.

3. Rank managers based on how much they emphasize shared goals, emphasize teamwork, permit autonomy, and give rewards in their leadership style.

4. Using your judgment and the leadership factor matrix, decide whether a manager places too much or too little emphasis on specific factors. Use this information in performance appraisals, training, or self-improvement.

Tool 99

Leadership Factor Matrix

Emphasis / Factors	TOO LITTLE	JUST RIGHT	TOO MUCH
SHARES GOALS	Dictates work	Shares ideas and goal setting with workers	Depends on workers to define tasks
EMPHASIZES TEAMWORK	No group meetings	Goals, assignments, and problem solving in group meetings	Too many group meetings
PERMITS AUTONOMY	Gives workers little freedom or responsibility	Allows workers freedom within established guidelines	Workers confused— no direction from manager
GIVES REWARDS	Criticizes, seldom compliments	Recognizes and demands good work, gives positive feedback	Rewards to gain acceptance, not in response to good work

Tool 100. Power Matrix

This method was developed by two researchers at Stanford University to help them understand how decisions are made at microcomputer companies. Decision making does not always follow departmental lines, and it is often unclear, even to a company's managers, exactly how decisions are made. The power matrix uses a short survey to rate each manager's influence over decisions. Influence in specific decision areas is also averaged to create a "power score" for each manager. The power scores can be used to compare the influence of individual managers, to find out which departments have the greatest influence in a company, to compare an organization chart (included as Tool 62) with actual distributions of power, or to find out which decision areas would be most affected by a proposed transfer or promotion.

In some cases, an organization does not really want to see data about the power of individual managers. However, this information is surprisingly useful when diagnosing problems, considering changes in organizational structure, or making promotion decisions. It is also useful in strategic planning, where it can be adapted to rank the importance of departments or strategic business units (SBUs), rather than managers, in decision making. Because the information may be controversial, chief executives should consider keeping the summary table confidential.

Here are the five steps to follow when using the power matrix.

1. Identify the key decision areas in a company. Start with the functional areas such as finance, manufacturing, and sales suggested by the organization chart and break these areas down into two or more areas if they represent multiple functions. For example, the marketing function might include advertising, new-product development, and customer research. Finance can be divided into external financing, funds management, and corporate expenditure approvals.

2. Turn the list of decision areas into a questionnaire by building a table with one column headed "Importance to long-run health of company" and additional columns labeled with the names or titles of managers. Head this section "Amount of influence manager has on decisions." Instructions should specify a 1-to-10 scale (1 equals unimportant/little influence and 10 equals very important/very strong influence.)

3. Each manager, including the chief executive officer, completes the questionnaire independently (but not anonymously). Managers

evaluate the influence of other managers but *not* their own influence.

4. Calculate a weighted-average power score for each manager and decision area. First, multiply each influence rating by the importance rating for the relevant decision area. These are the weighted influence scores. Second, compute the mean of all weighted influence scores for each manager.

   Example: Manager A gives advertising a 5 for importance and gives manager X a 6 for influence over advertising decisions. The weighted score is 5 x 6 = 30. X's weighted influence scores for advertising from the other managers are 28, 30, and 42. The four weighted influence scores total 130. The mean is 32.5, which is X's power score for advertising decisions.

5. Enter the mean power scores into a table with decision area down the left and managers across the top. (See Tool 100.) Sum each column and divide by the number of rows (or decision areas) to calculate the total power score for each manager.

   Idea: Calculate department scores by averaging managers' scores from each department. The department scores indicate which function drives the company. For example, in a company that is technology driven, the research and development (R&D) and manufacturing departments should have the highest power scores.

Tool 100 is a power matrix for a hypothetical computer company. The matrix shows a fairly even distribution of power—the spread between the highest and lowest total power scores is not great. In general, each manager has a strong influence over his or her functional area. However, note that the vice president of U.S. operations nearly has as strong an influence as the president over decisions concerning U.S. sales and marketing. The matrix indicates a possible problem area: Note that the vice president, Europe, has little input in R&D, purchasing, and financing decision areas—the organization is probably not considering the needs of its European markets when making product development decisions.

Use the power matrix to:

- Rank individuals, positions, or departments by the amount of decision-making power they have in an organization.

- Identify power centers for a reorganization or organization development study.

- Decide what an organization's functional orientation is for strategic planning purposes (i.e., operations-oriented vs. market-oriented).
- Update or redesign the formal organization chart to better reflect informal decision-making patterns.
- Evaluate the impact of a proposed management transfer or promotion.

As a recap, here are the five steps to follow when using the power matrix.

1. Identify and list the major decision areas in the organization.

2. Create a table of decision areas and decision makers.

3. Distribute the table as a questionnaire to managers. Instruct them to rate the importance of each decision area and the influence of every other manager on that decision area.

4. Calculate decision scores for each manager by averaging influence ratings and weighting these by their importance for each decision.

5. Create a power matrix showing individual manager's influence and total power scores.

Tool 100

Power Matrix

Decision area	President	VP Mfg	VP U.S. Ops	VP Europe	VP Finance	VP R&D
U.S. Sales and Marketing	8.3	2.8	8.2	6.3	2.7	3.5
European Sales & Marketing	4.7	2.3	7.8	9.1	2.5	3.2
R&D	5.8	5.6	4.5	2.5	3.7	9.3
Manufacturing	5.3	9.2	5.7	3.2	3.2	6.5
Purchasing	7.0	8.1	5.4	2.6	5.0	5.0
Financing	5.4	3.5	2.6	2.4	5.5	2.6
Total Power Score	6.1	5.3	5.7	4.4	3.8	5.0

Tool 101. Territory Sales Performance Chart

Bindicator, a manufacturer of instruments for measuring raw material levels in bins, needed better measures of sales territory performance in order to improve its marketing. Its nationwide sales force operated out of approximately 30 territories, and management wanted a better way to compare territory performance and allocate resources. The method the company used was developed in cooperation with Market Statistics and used Market Statistics' database. But this method can be applied successfully in-house by many organizations using other sources of data.

The following instructions use readily available census data (which, however, is not kept as current as Market Statistics' data). See *County Business Patterns* published by the Bureau of the Census, U.S. Department of Commerce, for each state of interest.

Here are the seven steps to follow when using the territory sales performance chart.

1. Assign Standard Industrial Classification system (SIC) codes to each customer in order to determine the percent of sales from each industry group. (*County Business Patterns* uses three-digit codes, so do not bother to break sales down by five-digit codes if this is your source.) Aggregate the sales figures for the last few years if there is significant year-to-year variation.

   Shortcut: If your company has many small customers, then analyze a random sample of accounts and assume that the industry composition of the sample applies to all sales. Make sure the sample includes all areas, as industry representation probably will vary with geographic location.

2. List the counties or states that compose sales territories. This may require some map work if you do not normally define territories by county or state borders.

3. Select a statistic that best represents sales potential from those available. In *County Business Patterns*, you have a choice of total payroll, total number of employees, number of businesses, and number of businesses by size based on number of employees. Counts of businesses are provided for each of these categories, by SIC code, county, and state. For example, an organization like Bindicator, which sells to large manufacturing and processing companies, might select the number of establishments with over 50 employees in designated SIC codes as their measure of sales potential.

4. Sum the chosen statistic for the appropriate SIC codes in each territory. Total the territory figures by SIC code, weight each SIC code by its percent contribution to your sales, and then sum across all SIC codes. When you have done this for each territory, find the total for all territories, and then calculate each territory's percent of the total.

5. Using actual sales figures, calculate each territory's percent contribution to your company's total sales. Use last period's sales figures, a three-year average, or the most recent figures from a trend line fitted to several years of sales data for each territory. (The latter is more difficult but preferable when there is consistent growth within territories and growth rates differ significantly between territories.)

6. For each territory, divide the percent of actual sales by the percent of expected sales. A negative number indicates that the territory's performance is below average, and a positive number suggests that it is above average. (**Warning:** The results are only as good as the statistics used to represent market potential.)

7. Then plot the data on a graph with the horizontal axis representing percent of expected sales and the vertical axis representing percent of actual sales. (See Tool 101A.) The action line is a diagonal line that shows average performance. Deviations from the diagonal action line indicate overperforming or underperforming territories. A logarithmic scale (included as Tool 51) on each axis makes this graph clearer.

8. ***Optional step for sales leads:*** Bindicator also plots percent of sales leads per territory vs. percent of expected sales to see whether lead-generating advertising is covering territories evenly. (See Tool 101B.) It found that some underperforming territories did not have adequate sales support. These are the action dots. (This is not surprising since advertising is usually allocated based on sales history.)

Use the territory sales performance chart to:

• Develop an objective measure by which to compare the performance of sales territories.

• Set sales goals based on the potential of a territory rather than on last year's performance in that territory.

• Evaluate the coverage of lead-generating advertising by territory.

Tool 101A

Territory Sales Performance *(vs. Potential)*

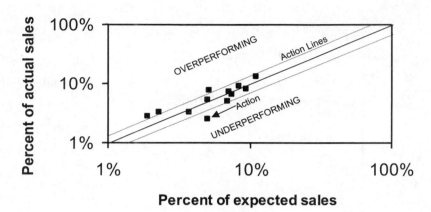

Tool 101B

Sales Leads per Territory *(vs. Potential)*

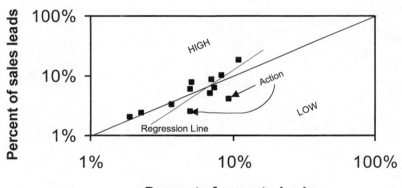

As a recap, here are the seven procedures to follow when using the territory sales performance chart.

1. Break down your company's sales by SIC code.

2. Define territories by the states or counties within them.

3. Choose a statistic to represent the potential market size for each territory.

4. Calculate total market potential per territory.

5. Calculate the contribution of each territory to your company's sales.

6. Calculate the ratio between actual sales and potential contribution for each territory.

7. Plot the data on a graph.

8. *Optional:* Calculate sales leads vs. sales potential per territory and plot this on a second graph.

16. A Thousand Points of Data

The obvious is that which is never seen until someone expresses it simply.

—Kahlil Gibran

Ever since the Big Bang, the universe has been expanding, and billions of new stars have been forming from the dust. The number of things to know is increasing much faster than our ability to comprehend them. That is true of business too. Businesses are generating data faster than any manager can master it. At the same time, the useful life of that data is collapsing. Corporations that used to develop an annual five-year strategic plan now revise their plans every few months. Lack of data is rarely a problem for a business. Rather, the problem is to refine data and present it coherently.

Corporations all around the world have access to almost unlimited amounts of data—for example, corporate databases and other sources contain data on sales, demographics, trends, competitive products, and so forth. Yet most corporations cannot effectively convert this data to useful information. How many corporate managers, for example, can easily determine the sales level of one of their products in a particular region? Can they access this information with a few mouse clicks, or does gathering and presenting the information require several people working for hours or even days? Such data certainly exists somewhere (most likely in a database); however, the speed at which the data can be converted to useful information depends on the information system and the graphical formats that the company uses.

Companies with systems and graphics that allow employees to communicate and analyze information rapidly enjoy a competitive advantage in the marketplace. When you use powerful graphics, you give viewers a greater understanding of information and increase the likelihood that they will make better decisions.

A poor presentation contains too much, too little, or disorganized data. A rich presentation often combines words, numbers, and graphics. Good graphics speed up comprehension. Every second counts, and saving time is one of the best corporate contributions and career investments that you can make.

Here are some other details to keep in mind when planning to incorporate graphics into your presentations and meetings.

- Use graphics, but come prepared with all the facts and figures as well. Graphics will not make up for a poorly prepared or boring presentation.

- Analyze the details of your presentation and rehearse before the meeting.

- Give the audience handouts of the charts in your presentation. Make sure your contact information is on them in case anyone has questions.

- Be sure the graphics you choose are appropriate to your audience and topic.

- To explain complex ideas or data, use the "PGP" method: Particular/General/Particular. For example, introduce your charts by explaining one part in detail, linking it to a general discussion of what the data means. Then return to the detail.

- Do not saddle your graphics with too much detail.

- Double-check your graphics for factual errors.

- Show up early to troubleshoot any technical problems with overhead projectors, laptop computers, or other presentation devices.

- When graphics are essential to your presentation, carry them on the plane with you. Your slides will do you absolutely no good if they wind up in Cleveland when your presentation is in Raleigh.

There is a difference between thinking through the use of charts and graphs consciously and actively, and merely operating on autopilot. Do not just be on autopilot. No single tool will ever meet all the needs of every enterprise, and the tools must be customized to reflect your organization's unique business requirements and characteristics. If you use these tools imaginatively, you will find that imagination will become your strongest tool. Imagination is best defined as the ability to see ordinary things in new ways, but even that does not quite get at the fundamental imaginative activity: putting old ideas together in new ways. For instance, John Lennon took the opening chords of Beethoven's Moonlight Sonata, rearranged them, and created the song "Because."

In business, originality can lead to success, but success also can be the result of "borrowing" great ideas from somewhere else. Brilliance in business means knowing when to put old ideas together in new ways and how to apply breakthroughs from other industries to your own. For example, one summer night shortly before the outbreak of World War I, Duncan Black and Alonzo Decker were sitting at a kitchen table. On the table was a Colt revolver. That night they would apply the fundamental design concepts of the revolver—a pistol grip and a trigger switch—to the electric drill. Their brainstorm resulted in the birth of the modern power tools industry.

Another example of this type of imagination occurred approximately 70 years later, when Robert Crandall masterminded an ingenious way to foster customer loyalty with American Airline's frequent-flyer program. His "original" idea was borrowed from somewhere else: namely, S&H Green Stamps. This idea continues to prove itself transferable, and it has gone on to inspire loyalty programs in businesses ranging from bakeries to ski resorts.

In a similar fashion, good charts can be the result of applying techniques that have been used elsewhere and tailoring them to the unique characteristics of your organization and situation. Business is complex, and no one can spend his or her career entirely creating charts. Many things besides charts and graphs matter. There is obviously a lot more to running a company. Charting a business issue, however, is also a complex and difficult activity. Even though all business issues do not lend themselves to resolution just because of good charts, poor charts can certainly confuse an already complex situation. If you use charts and graphs to improve your presentations and reports, then you can eliminate unnecessary static in the give-and-take between you and your business colleagues.

The need for good charts and graphs is so obvious that it is hard for anyone to be against them. "Good design," says Edward Tufte, "is clear thinking made visible." In oral presentations and written ones, good charts and graphs encourage people to want to know more about your subject. Charts and graphs are forms of communication that make their way easily across barriers of language and culture. They are objective, have few cultural or regional differences, and can be understood by anyone. They are intriguing and pique curiosity.

Ultimately, the value of charting to managers lies as much in the process of creating the chart as in the consumption of its product. Charting stimulates the development of a deeper understanding of a business and its environment, and it forces the formulation and evaluation of alternatives that would not otherwise be considered. It unleashes large amounts of creativity that are so often suppressed by routine and the need to respond to crises.

Strive for an awareness of the best tools that are available. Use these tools to improve your analysis and presentation of financial and management information. Keep your toolbox well stocked, and always reach for the simplest and clearest techniques. If you improve the way that you communicate, then you cannot avoid improving the profitability of your business.

Appendix A.
Anatomy of a Chart

To understand how a chart can provide precise information, we must look closely at its structure. All charts, no matter how different individual examples may look, are created from the same components. Typically, they have three primary elements: *framework, content,* and *labels.* Thus stripped down, Chart A.1 is similar to the first chart we met in high school, with its calibrated vertical Y axis and horizontal X axis and origin at 0. In algebra class, the content of such a chart was often the curve produced by plotting an equation—a special case of a relationship among numbers.

The framework of the chart sets the stage, indicating what kinds of measurements are being used and what things are being measured. The simplest framework has an L shape, one leg standing for the amount of a measured substance and the other for the things being measured. The vertical leg, the Y axis, usually stands for the measurements (dollars, barrels of oil, degrees of temperature), and the horizontal leg, the X axis, for the things being measured (products, years, territories).

The content is the lines, bars, point symbols, or other marks that specify particular relations among the things represented by the framework. The positions of content elements are plotted as values along the Y axis (e.g., dollars) and are paired with values along the X axis (e.g., months).

Each leg of the framework bears a label naming a dependent variable (the type of measurement being made) or an independent variable (the entity to which the measurement applies). The title of the chart is itself a kind of label.

Charts also may include a number of optional components. For example, some charts may include inner grid lines. These lines stretch across the framework at regular intervals, either horizontally, vertically, or in both directions.

Charts sometimes include a key takeaway box. It is a comment on or summary of the display, a short description that explains key terms or directs the reader's attention to specific features of the display.

Chart A.2 shows the anatomy of a chart in more detail. It contains the following components.

A	Y-axis title	G	Gridline
B	Value (Y) axis	H	Tick mark
C	Data label	I	X-axis title
D	Chart title	J	Tick-mark label
E	Data marker	K	Category (X) axis
F	Legend		

These are the chart options that are available when doing financial charting in Excel 97.

Chart A.1

Anatomy of a Chart

Title

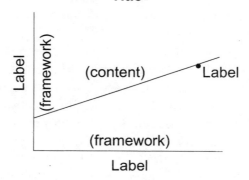

The primary elements of a chart: framework, content, and labels. This is a key takeaway box, which explains key terms, summarizes, or suggests recommended action.

Chart A.2

Anatomy of a Chart—Detail

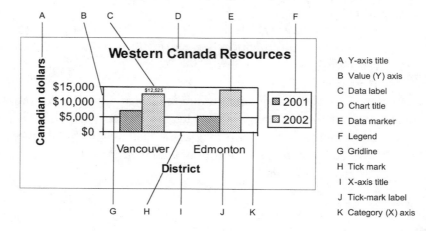

A Y-axis title
B Value (Y) axis
C Data label
D Chart title
E Data marker
F Legend
G Gridline
H Tick mark
I X-axis title
J Tick-mark label
K Category (X) axis

Appendix B.
Defining Chart Types

Each chart type offers special advantages for analyzing data and presenting numerical information. When considering the type of chart to choose, keep in mind how you will present the chart. If you plan to use it in a presentation, keep it as simple as possible. (During a presentation, your audience may not have time to interpret a chart that contains many series.) If the chart is going into a written report, then it can contain more information, because your readers can take their time to analyze the contents.

When creating charts in Excel 97, the following sections describe chart types.

Area Chart Type

An area chart emphasizes the magnitude of change over time. By displaying the sum of the plotted values, an area chart also shows the relationship of parts to a whole.

Bar Chart Type

A bar chart illustrates comparisons between individual items. Categories are organized vertically, values horizontally, to focus on comparing values and to place less emphasis on time.

Stacked bar charts show the relationship of individual items to the whole.

A three-dimensional (3-D) bar chart emphasizes the values of individual items at specific times or draws comparisons between items. The subtypes stacked and 100 percent stacked bar charts show relationships to a whole.

Column Chart Type

A column chart shows data changes over a period of time or illustrates comparisons between items. Categories are organized horizontally, values vertically, to emphasize variation over time.

Stacked column charts show the relationship of individual items to the whole.

A 3-D column chart shows a 3-D view of a column chart in one of two variations: simple 3-D and 3-D perspective. The simple 3-D column displays the column markers along the X (or category) axis. The 3-D perspective chart compares data points along two axes: the X axis and the Y (or value) axis. In both chart variations, the data series are plotted along the Z axis. This chart type allows you to compare data within a data series more easily and still be able to review the data by category.

Line Chart Type

A line chart shows trends in data at equal intervals. Although line charts are similar to area charts, line charts emphasize time flow and the rate of change rather than the amount of change or the magnitude of values. A 3-D line chart shows a 3-D view of a line chart as 3-D ribbons. This chart type is often used to display data attractively for presentations.

Pie Chart Type

A pie chart shows the size of items that make up a data series proportional to the sum of the items. It always shows only one data series and is useful when you want to emphasize a significant element. To make small slices easier to see, you can group them together as one item in a pie chart and then break down that item in a smaller pie or bar chart next to the main chart.

XY (Scatter) Chart Type

An XY (scatter) chart either shows the relationships between the numeric values in several data series or plots two groups of numbers as one series of XY coordinates. It shows uneven intervals, or clusters, of data and is commonly used for scientific data. When you arrange your data, place X values in one row or column, and then enter corresponding Y values in the adjacent rows or columns.

Doughnut Chart Type

Like a pie chart, a doughnut chart shows the relationship of parts to a whole, but it can contain more than one data series. Each ring of the doughnut chart represents a data series.

Radar Chart Type

In a radar chart, each category has its own value axis radiating from the center point. Lines connect all the values in the same series. A radar chart compares the aggregate values of a number of data series.

Surface Chart Type

A surface chart is useful when you want to find optimum combinations between two sets of data. As in topographic maps, colors and patterns indicate areas that are in the same range of values. A 3-D surface chart shows a 3-D view of what appears to be a rubber sheet stretched over a 3-D column chart. Surface charts can be used to show relationships between large amounts of data that otherwise may be difficult to see.

The colors do not mark the data series. The wire-frame format displays that data in black and white.

The contour chart formats provide a two-dimensional (2-D) view of the data from above, similar to a 2-D topographic map.

Bubble Chart Type

A bubble chart is a type of XY (scatter) chart. The size of the data marker indicates the value of a third variable. To arrange your data, place the X values in one row or column, and enter corresponding Y values and bubble sizes in the adjacent rows or columns.

Stock Chart Type

The stock chart is used often to illustrate stock prices. This chart can be used for scientific data, for example, to indicate temperature changes. You must organize your data in the correct order to create stock charts. A stock chart that measures volume has two value axes: one for the columns that measure volume, the other for the stock price. You can include volume in a high-low-close or open-high-low-close chart.

Cone, Cylinder, and Pyramid Chart Types

The cone, cylinder, and pyramid data markers can lend a dramatic effect to 3-D column and bar charts.

Appendix C.
Financial Charting in Excel 97

To create a chart, you must first have a range of data. Once you have your data, follow these three steps:

1. Select the entire range of data to be charted.

2. From the Insert menu, choose the Chart command. (You also can click the Chart Wizard button on the Standard toolbar.) This launches Excel's Chart Wizard, which steps you through four dialog boxes that help you to design a chart.

3. Specify the various elements of your chart (i.e., type, format, data series, axes, legends, chart title, and axis title) in the first three dialog boxes. The fourth dialog box allows you to select a location for your chart.

The following four steps are the four dialog boxes in Chart Wizard. (Note that the optional boxes used in each step of the Chart Wizard are also available again through the right mouse menu.)

Step 1 of 4—Chart Type

Standard Charts: This feature allows you to select from a list of standard chart types.

Custom Types: The custom types option allows you to select from a list of combination, specialty, and user-defined charts.

Preview: The "Press and hold to preview sample" button allows you to determine if the chart is appropriate.

Step 2 of 4—Source Data

The Source Data dialog allows you to select the cells or range to chart. If the chart data range was selected when you started the Chart Wizard, then the cells selected will appear in the dialog. If you want to choose another data range, click on the button

(Collapse Dialog Button) to the right of the data range field and reselect the data.

Step 3 of 4—Chart Options

Chart Options allow you to specify titles for the chart and the axes, set the primary axis, show or hide gridlines, specify the placement for the legend, select the type of data label, and show or hide the data table.

Step 4 of 4—Chart Location

Chart Location asks you if you want to place the chart on a new sheet or as an object in a worksheet. This allows you to specify a location for your chart. If you select new sheet, a new chart worksheet is created and named. If you select object in, the chart is inserted into a worksheet as an object.

After it has been created, here is how to modify the chart.

• Point to each object for a tool tip identifying that chart object.

• Modify each part of the chart by double-clicking on it.

• Right-click each chart object for the mouse menu.

References

Chapter 1

Geri McArdle, *Delivering Effective Training Sessions for Productivity (50 Minute Series)* (Menlo Park, CA: Crisp Publications, 1994).

Edward R. Tufte, *Visual Explanations: Images and Quantities, Evidence and Narrative* (Chesire, CT: Graphics Press, 1997).

Chapter 4

Continuing Process Control and Process Capability Improvement: A Guide to the Use of Control Charts for Improving Quality and Productivity for Company, Supplier and Dealer Activities (Statistical Methods Office, Operations Support Staff, Ford Motor Company, September 1985).

L. P. Sullivan, "Reducing Variability: A New Approach to Quality," *Quality Progress* (July 1984), pp. 55–59.

Donald J. Wheeler, *Understanding Variation: The Key to Managing Chaos* (Knoxville, TN: SPC Press, 1993).

Chapter 5

Elwood S. Buffa, *Meeting the Competitive Challenge: Manufacturing Strategy for U.S. Companies* (Homewood, IL: Dow Jones-Irwin, 1984).

Robert D. Buzzell and Bradley T. Gale, *The PIMS (Profit Impact of Market Strategy) Principles: Linking Strategy to Performance* (New York, NY: The Free Press, 1987).

"Internet Trip Time Planner," *Computers in Physics* 8(2) (March/April 1994), p. 147.

Karou Ishikawa, *What Is Total Quality Control? The Japanese Way*, trans. David J. Lu (Englewood Cliffs, NJ: Prentice-Hall, 1985).

Edwin Mansfield, *Economics: Principles, Problems, Decisions*, 2nd ed. (New York, NY: W.W. Norton & Company, 1977).

Chapter 6

Herman Chernoff, "The Use of Faces to Represent Points in k-Dimensional Space Graphically," *Journal of the American Statistical Association* 68 (June 1973), pp. 361–368.

Chapter 8

"Most Admired Corporations," *Fortune* (March 2, 1998).

Chapter 9

Kenneth J. Albert, *How to Solve Business Problems* (New York, NY: McGraw-Hill, 1978).

H. Igor Ansoff, *Implanting Strategic Management* (Englewood Cliffs, NJ: Prentice-Hall, 1984).

William R. King, "Using Strategic Issue Analysis," *Long Range Planning* 15(4) (1982), pp. 33–48.

Milton D. Rosenau, Jr., *Innovation: Managing the Development of Profitable New Products* (New York, NY: Van Nostrand Reinhold, 1982).

Chapter 10

Raymond Villers, *Research and Development: Planning and Control* (New York, NY: Financial Executives Research Foundation, 1964).

Chapter 11

David Anderson, Dennis Sweeney, and Thomas Williams, *Quantitative Methods for Business,* 2nd ed. (St. Paul, MN: West, 1983).

Joseph Conrad, *Lord Jim* (New York, NY: Penguin Books, 1949).

Charles A. Holloway, *Decision Making Under Uncertainty: Models and Choices* (Englewood Cliffs, NJ: Prentice-Hall, 1979).

Donald D. Lee, *Industrial Research: Techniques and Practices,* 2nd ed. (New York, NY: Van Nostrand Reinhold, 1984).

Howard Raiffa, *Decision Analysis: Introductory Lectures on Choices under Uncertainty* (Reading, MA: Addison-Wesley, 1968).

Chapter 12

Derek F. Abell and John S. Hammond, *Strategic Market Planning* (Englewood Cliffs, NJ: Prentice-Hall, 1979).

Francis Aguilar, "The Mead Corporation: Strategic Planning," Harvard Business School Case 9-379-070.

Noel Capon, "Product Life Cycle." In Benson P. Shapiro, Robert J. Dolan, and John A. Quelch, *Marketing Management: Strategy, Planning and Implementation,* vol 2. (Homewood, IL: Richard D. Irwin, Inc., 1985).

Arnoldo C. Hax and Nicols S. Majiluf, *Strategic Management: An Integrative Perspective* (Englewood Cliffs, NJ: Prentice-Hall, 1984).

Arnoldo Hax and Nicolas S. Majiluf, *Strategic Management* (Englewood Cliffs, NJ: Prentice-Hall, 1984).

Bruce Henderson, *The Logic of Business Strategy* (Cambridge, MA: Ballinger, 1984).

Philip Kotler, *Marketing Management: Analysis, Planning and Control,* 5th ed. (Englewood Cliffs, NJ: Prentice-Hall, 1984).

Michael Porter, *Competitive Advantage: Creating and Sustaining Superior Performance* (New York, NY: Free Press, 1985).

Royal Dutch Shell Company, *The Directional Policy Matrix: A New Aid to Corporate Planning* (Royal Dutch Shell Company, 1975).

William E. Rothschild, *Strategic Alternatives: Selection, Development and Implementation* (New York, NY: Amacom, 1979).

William E. Rothschild, *Strategic Thinking* (New York, NY: Amacom, 1976).

Benson P. Shapiro, V. Kasturi Rangan, Rowland T. Moriarty, and Elliot B. Ross, "Manage Customers for Profits (Not Just Sales)," *Harvard Business Review* (September-October 1987), pp.101–108.

Chapter 13

Keith R. Herrmann, "The One Page CFO," *Management Accounting* (February 1995), pp. 55–59.

Robert Rachlin, *Return on Investment Manual: Tools and Applications for Managing Financial Results* (Armonk, NY: Sharpe Professional, 1997).

Chapter 14

"Amana's Analysis," *Management Briefing: Marketing* (New York, NY: The Conference Board, April 1987).

H. Igor Ansoff, *Implanting Strategic Management* (Englewood Cliffs, NJ: Prentice-Hall, 1984).

David W. Cravens, "Gaining Strategic Marketing Advantage," *Business Horizons* (September-October 1988), pp. 44–54.

Philip Kotler, *Marketing Management: Analysis, Planning, and Control,* 5th ed. (Englewood Cliffs, NJ: Prentice-Hall, 1984).

Otto Ottesen, "The Response Function." In *Current Theories in Scandinavian Mass Communications Research,* ed. Mie Beg (Grenaa, Denmark: G.M.T., 1977).

Chapter 15

Karl Albrecht, *Organization Development* (Englewood Cliffs, NJ: Prentice-Hall, 1983).

Robert R. Blake and Jane S. Mouton, *The Grid for Sales Excellence: New Insights Into a Proven System of Effective Sales,* 2nd ed. (Austin, TX: Scientific Methods, Inc., 1970).

L. J. Bourgeois III and Kathleen M. Eisenhardt, "Strategic Decision Processes in High Velocity Environments: Four Cases in the Microcomputer Industry," *Management Science* 34 (7) (1988), pp. 816–835.

Karsten Hellebust, "Bindicator Finds a Fair Measure of Sales Territory Performance," *Sales and Marketing Management* (November 11, 1985), pp. 86–93.

Management Briefing: Marketing (New York, NY: The Conference Board, October 1987).

Index

For information about the CD-ROM, see the **About the CD-ROM** section on page xix.